J. NORMAN LOCKYER

THE

SUN'S PLACE IN NATURE

Elibron Classics
www.elibron.com

Elibron Classics series.

ISBN 0-543-99967-X (paperback)
ISBN 0-543-99966-1 (hardcover)

This Elibron Classics Replica Edition is an unabridged facsimile
of the edition published in 1897 by Macmillan and Co., Ltd., London.

THE SUN'S PLACE IN NATURE.

THE

SUN'S PLACE IN NATURE

BY

SIR NORMAN LOCKYER, K.C.B., F.R.S.

Correspondant OF THE INSTITUTE OF FRANCE; THE SOCIETY FOR THE PROMOTION OF NATIONAL INDUSTRY OF FRANCE; THE ROYAL ACADEMY OF SCIENCE, GÖTTINGEN; THE FRANKLIN INSTITUTE, PHILADELPHIA; THE ROYAL MEDICAL SOCIETY OF BRUSSELS; SOCIETY OF ITALIAN SPECTROSCOPISTS; THE ROYAL ACADEMY OF PALERMO; THE NATURAL HISTORY SOCIETY OF GENEVA; *Member* OF THE ROYAL ACADEMY OF LYNCEI, ROME; AND THE AMERICAN PHILOSOPHICAL SOCIETY, PHILADELPHIA; *Honorary Member* OF THE ACADEMY OF NATURAL SCIENCE OF CANTANIA; PHILOSOPHICAL SOCIETY OF YORK; LITERARY AND PHILOSOPHICAL SOCIETY OF MANCHESTER; AND LEHIGH UNIVERSITY; *Director* OF THE SOLAR PHYSICS OBSERVATORY; AND *Professor* OF ASTRONOMICAL PHYSICS IN THE ROYAL COLLEGE OF SCIENCE, LONDON.

London

MACMILLAN AND CO., LIMITED

NEW YORK: THE MACMILLAN COMPANY

1897

LONDON:
HARRISON AND SONS, PRINTERS IN ORDINARY TO HER MAJESTY,
ST. MARTIN'S LANE.

PREFACE.

Among the conclusions arrived at, as a result of the general inquiry referred to in the introductory chapter, it was not to be wondered that many were entirely novel, and, this being so, a very close criticism of them was to be expected, and, indeed to be hoped for. In the preface to the work, entitled *The Meteoritic Hypothesis*, in which I gave a general account of the conclusions, I wrote:—

"It is not in the nature of things that a large mass of detailed work and inquiry, which it has taken my assistants and myself three years to get together, shall be found free from error, especially since observations made by many men in many lands, frequently under conditions of great difficulty, form part of the basis of the discussion. Nor, again, is it likely, or even desirable, that the general hypothesis, if it be found of any value at all, shall not be improved when fresh minds are brought to bear upon it.

"When the time arrives I shall profit more than any one else by any valid objections that may be raised, and I shall be careful to reply to or accept them."

In the year 1894 my turn came round, as Professor in the Royal College of Science, to give one of the courses of "Lectures to Working Men" at the School of Mines, and it struck me that I could not do better than give a general account of the bearing of the new conclusions upon solar studies, for, in fact, the whole inquiry had been originated by the desire to obtain definite evidence as to the relation of the sun to the more distant stars. And I was the more induced to do this because it would afford me a convenient opportunity of bringing together all the new knowledge which the four years since the publication of *The Meteoritic Hypothesis* had brought, and of discussing whether this, together with the remarks and criticisms which had accompanied it, had strengthened or weakened the views I had put forward.

The lectures had for a title, "The Sun's place in Nature," and the present work is to some extent based upon the shorthand notes of them.

I have endeavoured to give in this book, as clearly and as judicially as I can, a statement of the discussions which have been going on since *The Meteoritic Hypothesis* was published; to show what holes have been picked in the new views, and what new truths may be gathered from the new work which has now been brought to bear upon the old; so that, as a result, the place I have given to the sun among its fellow stars may be justified or withdrawn.

To the subject matter of the lectures many additions have been made in the light of more recent work. The running to earth of the cause of the D_3 line—which, in 1868, I attributed to a gas associated with hydrogen, and named helium—has an important bearing on many of the points discussed; hence special chapters have been devoted to it.

The outburst of a new star in the constellation of Auriga in 1895 I at once accepted as a test of the validity of the new hypothesis, and I have therefore referred at length to the question of new stars generally and to the conclusions which have been drawn from the study of the last one by modern methods.

Finally, I give a general account of several researches I have quite recently communicated to the Royal Society dealing especially with stellar classification.

I have been careful, I trust, to reply to the objections which have been made to my views, and also to show the bearing of the new work upon them. In my attempt to do this thoroughly I have had to treat on various problems of celestial chemistry which are now rapidly being opened up.

I have to acknowledge my deep obligations to the staffs both of the Solar Physics Observatory and of the Astrophysical Department of the Royal College of Science, chief among them Messrs. Fowler, Baxandall, and Shackleton, for the skill and patience with which they have assisted me in the various branches of the inquiries summarised in this volume, which have occupied us now for ten years.

NORMAN LOCKYER.

Solar Physics Observatory,
South Kensington.
July, 1897.

CONTENTS.

BOOK I.—INTRODUCTORY.

BOOK II.—NEBULÆ AND STARS.

BOOK III.—NEW STARS.

BOOK IV.—THE SUN'S PLACE AMONG THE STARS.

LIST OF ILLUSTRATIONS.

THE

SUN'S PLACE IN NATURE.

CHAPTER I.—INTRODUCTORY.

In the year 1887 I brought out a book entitled *The Chemistry of the Sun*, the subject matter of which had grown out of work upon which I had then been engaged for twenty-one years, the latter part of it having been done in connection with a Government Committee—the Solar Physics Committee. The object of the work was to endeavour to obtain some light upon the nature of the sun, among other things with a view of determining whether a complete investigation of the quantity and the quality of the solar radiation might, sooner or later, not only provide us with new knowledge, but with some help for humanity in the way of long period weather prediction.

It is no part of my present intention to dwell at any length on the older inquiries made in this connection, but I must refer to two points on which the work to which I have referred certainly threw very great interest, if not light.

The radiant energy of the sun, measured in a mechanical way, is, according to Lord Kelvin, equal to 78,000 horse-power per square metre, a square metre being a little more than a square yard. That means, therefore, that from every space on the sun's surface, a little bigger than a square yard, enough energy is produced to drive three or four of our biggest iron-clads.

A later value, found by Langley, gives 135,000 instead of

78,000 horse-power per square metre. What, then, is the effective temperature of the sun's photosphere which causes it to give out all this energy? Recent investigations by Rosetti, Le Chatelier, Wilson and Gray indicate that a value of 9000° C. is not very wide of the mark.

Now, one of the principal results obtained by spectroscopic work during the last thirty years is that spectra vary as the temperature varies; and, working on these lines, it was soon found that, on certain parts of the sun—which it is possible for us to study by modern means of observation—there was no temperature we could employ in our laboratories which exactly represented the heat apparently at work in these regions. Spectrum analysis, therefore, at most temperatures we could command in our laboratories, seemed to break down, because the whole soul of spectrum analysis, as applied to celestial investigation, means that we must match what we see shining in the heavens by something that we can make shine in our laboratories under equivalent temperature conditions. Nor was this all. The more the researches went on, the more the spectroscopic observations of sun spots and the like seemed to suggest that the meteoric origin of the sun, and of some part of the sun's radiant energy, suggested many years ago, was a true one; it certainly appeared to explain a great many solar phenomena, which have not been explained in any other way.

On this point I wrote in 1886:—

"We know that small meteorites in our own cold atmosphere are heated to incandescence by friction, that is, by the conversion of their kinetic energy into heat, and it is therefore not difficult to imagine that enormous masses, falling with great velocities through the sun's highly heated atmosphere, would be competent to give rise to such disturbances as those with which we are familiar on the sun's surface. This cool material is produced by the condensation, in the upper cool regions of the sun's atmosphere, of the hot ascending vapours produced at the lower levels, and this is probably the main source of supply of spot-producing material."[1]

---

[1] *Encyclopædia Britannica*, vol. xxii, art. "Sun."

It seemed, indeed, as if many phenomena on the nearest star to us, our own sun, might be really phenomena produced by the fall of meteoric bodies upon that surface which we see, and which we call the photosphere. It is now many years ago since Balfour Stewart and others threw out the idea that the phenomena connected with the formation of sun-spots were really produced by the fall of bodies upon the sun's surface. Other philosophers have preferred the idea that we have to do with eruptions from the interior of the sun; nothing can be more divergent than the opinions which have been brought forward as explanations of these appearances.

We at once see that, if we assume that this meteoritic action may take place in the solar atmosphere, it need not necessarily be, and, in all probability, is *not* a meteoritic action coming from without.

Taking our own case; we live in a damp climate, and sometimes the air is dampest when there are no clouds. Clouds are condensations of the moisture in the air, and we know that it is not really a question of clouds only; we may have snow, rain, or hail, and all these represent different condensations of the damp—or, as we call it, the aqueous—vapour which is ever present in our air. Apply that to the sun. What is the air of the sun composed of? Certainly one important constituent of it is the incandescent vapour of iron; we are no longer dealing with a low temperature and the vapour of water, but with an atmosphere in the hotter parts of which iron is not solid or liquid, but in which the temperature is high enough to keep it in a state of gas, probably thousands of degrees higher than is arrived at in the Bessemer process.

We will assume, then, that that temperature and that condition of atmosphere prevails for 20,000 miles above the photosphere of the sun. As we get further from the sun, the atmosphere is, of course, getting cooler, and at some distance above the photosphere the temperature will certainly be so reduced

B 2

that the iron vapour may play the part of our aqueous vapour; then it condenses and turns into iron snow and iron hail and iron rain, and so on, falling upon the photosphere as the rain falls on the earth. There is thus a possibility in the sun of home-made meteoritic action.

These conclusions raised a great many interesting questions. We had the sun, to judge by the divergence in the spectral phenomena, hotter than anything we could get at in our laboratories; if we accepted the prevailing notion of the constitution of other celestial bodies, we found in the nebulæ bodies which, according to the then received notions, were themselves so hot that the light which was radiated by them came from hydrogen or something finer than hydrogen, associated with nitrogen or something finer than nitrogen; in other words, the nebulæ were stated to be masses of gas hotter than the sun.

If then the sun, as representing the stars, is very hot and the nebulæ are very hot, where were the cooler bodies in space? It was therefore very important from the solar physics point of view that this question should be investigated, and to do this the first thing to be accomplished was to determine, if possible, the sun's place among the stars, and to learn more, if possible, concerning the stars themselves.

I determined to see how all the spectroscopic observations which had been made up to 1886 bore out a suggestion which had been made in 1871 by Professor Tait, before there was very much spectroscopic evidence to go upon. The result was that my assistants and myself spent something like three years in gathering together, we believed, every available observation; at all events, if not every available observation, there were between thirty and forty thousand of them, and we found that a very considerable number. I not only determined to collect them, but also to discuss them, and make any experiments or observations which might be suggested by the discussion. In order to effect this, the years 1887—1890 were spent in bringing together and co-ordinating the observations which had been made up to that time on the spectra of the various orders of cosmical bodies in connection with laboratory work.

The hypothesis to which I was led seemed to indicate the sun's place among the stars beyond all question, but, more important even than this, it suggested that meteoritic action was accountable for very much more than the production of many of the solar phenomena.

What I found was that when we discussed all the observations we could get together, and in relation to stars as well as nebulæ and comets, a meteoritic origin of the various classes of heavenly bodies seemed to explain many things, and threw a perfectly new light upon the visible universe; there were, moreover, several points raised of intense novelty and freshness, each of which could be discussed separately, cast aside if it were false, and held on to if it were true.

The following were the most important conclusions which I refer to in what I consider to be the order of their importance:—

1. There is the closest possible connection between nebulæ and stars. They represent two stages in an evolutionary series.

2. The first or nebulous stage in the development of cosmical bodies is not a mass of hot gas, but a swarm of cold meteorites.

3. Many bodies in space which look like stars are really centres of nebulæ: that is, of meteoritic swarms.

4. Stars with bright-line spectra must be associated with nebulæ.

5. Double swarms, in any stages of condensation, may give rise to the phenomena of variability.

6. New stars are produced by the clash of meteor swarms. They are closely related to nebulæ and bright-line stars.

7. Since on the meteoritic hypothesis some of the heavenly bodies must be increasing their temperature, while others are

decreasing it, a new classification of the heavenly bodies is demanded, based on the varying states of condensation of the meteoritic swarms.

The complete investigation gave as an answer to the question which had suggested it, that the sun is one of those stars the temperature of which is rapidly decreasing, and that many of the changing phenomena of the sun are due to the fall of meteoritic matter upon the photosphere.

There are few things more remarkable in the recent history of astronomical work than the enormous expansion of our knowledge on many of the above points within a very short period after the new views had been put forward.

Almost all the observations discussed in my book entitled *The Meteoritic Hypothesis* were made before the year 1888. That year really marks an epoch in astronomical observations, and the results achieved since that time are remarkable for their number and importance, so much so that the new work to be now discussed equals, in many lines of the inquiry which now concerns us, the volume of all preceding observations. Indeed, it would be very difficult to over-estimate the enormous advantages under which such work is now carried on; advantages in that now, when any question is put to any part of the heavens, we know that there are many good workers employed under the best possible conditions to get the particular information that we want recorded for the most part by photography.

It is not wonderful, then, that with such advantages in every branch of inquiry we find advances gigantic, marvellous, almost beyond belief.

Owing to this, it has been possible in subsequent discussions to replace most of the old observations (which by many I have been blamed for using, although they were the only ones available at the time), by others, in the securing of which all the resources of modern science and skilled investigation have been employed without stint. Not only is there much new matter to discuss, but the area of fact has been enormously widened, and this is all important when the testing of new views is in question.

Some of the most important points I propose, therefore, to discuss again in the light of the new work, but before I do so it will be well to dwell for a little on the recent appliances to which I have referred. It seems especially desirable to indicate the classes of instruments from which this new work has come —work which so overshadows the old that in only a few years the old observations will cease to be considered altogether when definite numerical statements have to be made.

CHAPTER II.—OUR PRESENT OBSERVING POWER.

WHEN we come to consider in detail the reason of the recent advance in observing powers, there can be no question that in the main it is due to the larger dimensions of the telescopes brought to bear upon physical inquiries.

I am sorry to say that with regard to the refractor the centre of gravity of the activity has left our country and has gone out West. We have to look to our American cousins for a great deal that we want to know in these matters, for the reason that now they not only have the biggest refractors and most skilled observers, but also they have been wiser than we—they have occupied high points on the earth's surface, and thus got rid of the atmospheric difficulties under which we suffer in England.

I may here refer to one of the most perfect pieces of workmanship in the world, constructed to investigate the phenomena of the heavens—the Lick Observatory, situated at an elevation of 4,000 feet on Mount Hamilton. Mr. Lick, the founder, has made his name immortal by helping on the progress of mankind. I wish some Englishman would immortalise himself in the same way. The principal instrument of this great observatory is a refracting telescope having an object-glass 3 feet in diameter, and a tube 56 feet in length. This is practically the most important telescope in the world at the present moment, and to give an idea of the wonderfully broad way in which the authorities have gone to work, I need only state the following fact. In an observatory it is sometimes difficult to get the observing chair at the right height or in the right position for observing a star or any celestial body with any comfort. The Americans get over this by simply raising the floor. By means of hydraulics the enormous floor, some 80 feet in diameter, is moved up and down with the chair.

Mount Hamilton, in the Northern Hemisphere, was re-echoed in 1891 by the Harvard College Observatory, at a height of 8,000 feet at Arequipa in the Southern Hemisphere, and not only so, but the 36-inch Lick telescope will soon be out-distanced by one of 40 inches aperture, also in the United States.

The Yerkes Observatory, which is to shelter this enormous refractor, was founded in 1892 by Mr. Charles T. Yerkes, of Chicago. It is similar to the 36-inch Lick telescope, but is heavier and more rigid, and many improvements have been introduced. An important feature, long ago suggested by Grubb and others, but apparently employed for the first time in this telescope, is a system of electric motors, by means of which the various motions, etc., are effected. The object-glass, by Clark, has recently been tested by Professor Keeler. The definition was found to be fully equal to that of the Lick telescope, while the light gathering power is considerably greater. The attachments of the Yerkes telescope will include—

1. A position micrometer by Warner and Swasey.

2. A solar spectrograph for micrometrical and photographic investigations of the spectra of solar phenomena.

3. A spectroheliograph for photographing the solar chromosphere, prominence, and faculæ by monochromatic light.

4. A stellar spectrograph for researches on the spectra and motions of stars, nebulæ, comets, and planets.

5. A photoheliograph of great focal length for photographing the direct solar image on a large scale.

The dome is 90 feet in diameter, allowing ample space for the tube of the great telescope, which, with its attachments, is about 75 feet long. The elevating floor of the observing room is 75 feet in diameter, and will be movable through a range of 22 feet by means of electric motors.

The instrument is now approaching completion. The accompanying figure will give an idea of its enormous dimensions.

FIG. 1.—Erecting the declination axis of the Yerkes 40-inch refractor.

But, thanks to the skill and generosity of Dr. Common, England's position with regard to reflectors has been vastly improved. Some of my own later work I shall have to refer to has been done by a 30-inch silver on glass mirror which he was kind enough to figure. Others, of various sizes up to 5 feet, are to be found at Ealing (his own observatory), the Solar Physics Observatory at Kensington, Greenwich, and Cambridge.

So much then for the increased supply of the astronomer's raw material—celestial light.

I have next to point out that the gradually lengthening exposures given to photographs of star-clusters and nebulæ has enormously extended our knowledge.

In this region of inquiry we have perhaps one of the most important instrumental advances, and the importance of it is not reduced by the fact that it was quite undreamt of when photographic processes were introduced. What we hoped to obtain were series of permanent and unbiassed records, but it did not strike us that the photographic processes could furnish us with records of phenomena which the eye has never seen and is never likely to see. Now that this benefit is known it is quite easy to explain it. No tyro using even the smallest camera but soon learns that the pictures he wishes to take may be easily over- or under-exposed, according to the time of exposure given. In the photographic process, therefore, he recognises that, unlike the case of the human eye, the longer an image is allowed to impress itself on the photographic plate, the brighter it becomes; in other words, very feeble luminous objects exposed for a considerable time will be recorded, and so on with increasingly long exposures, till at last a visible image is produced of an object utterly invisible to the eye. This then, shortly stated, is the origin of much modern work. Up to the present time the exposures have been increased to such an extent that thirty or forty hours is not out of the question. This means, of course, that the object, be it star or nebula, has to have a telescope pointed so perfectly to it that on as many successive nights as may be necessary, the images fall absolutely on the same part of the plate. It is obvious that this requires a perfection of pointing and adjustment which was undreamt of a few years ago.

One of the most important telescopes in the world at present is Dr. Roberts' reflector, with which long exposure representations of the heavenly bodies have been produced; these I shall have to refer to at one time or another in relation to different branches of our subject. In this instrument (Fig. 2) a reflecting telescope of 20 inches aperture is combined with a refractor of 7 inches aperture. The refractor is used as a guiding telescope, and ensures that the images of the stars and nebulæ fall on the same part of the photographic plate which is being exposed in the reflecting telescope throughout the whole

FIG. 2.—Dr. Roberts' twin telescope.

time of the exposure. Even with the best driving clocks, such a guiding telescope cannot be dispensed with when the exposures are prolonged for the number of hours necessary in some cases.

We next turn to the improved methods of obtaining the spectra of the various heavenly bodies.

I have gone so fully into the questions connected with spectrum analysis in a companion volume[1] that I need not enlarge upon it here, but by way of reminder I may in a few words show exactly the function of this new instrument of enormous power, which has in a very few years perfectly changed the aspect of astronomic science.

One key to the hieroglyphics, the light story, which is hidden in every ray of light, is supplied to us by the rainbow. It teaches us that the white light, which nature bountifully supplies us with in sunlight, is composed of rays of different kinds or of different colours; and it is common knowledge that there is an almost perfect analogy between these coloured lights and sounds of different pitches.

The blue of the rainbow may be likened to the higher notes of the key-board of a piano, and the red of the rainbow, on the other hand, may be likened to the longer sound waves, which produce the lower notes; and as we are able in the language of music to define each particular note, such as B flat and G sharp, and so on, so light-waves are defined by their colours or lengths.

What nature accomplishes by a rain drop we can do with a prism. If we pass a ray of white light through a prism, we find that after the light has so passed through, it is changed into a beautiful band, showing all the colours of the rainbow. Such a prism is the fundamental part of the instrument called the spectroscope, and the most complicated spectroscope which we can imagine simply utilises the part which this piece of triangular glass plays in breaking up a beam of white light into its constituent parts from the red to the violet. Between these colours we get that string of orange, yellow, green, and blue

---

[1] *Chemistry of the Sun*, Macmillan, 1887.

which we are familiar with in the rainbow. For sixpence any of us may make for ourselves an instrument which will serve many of the purposes of demonstrating some of the more beautiful fields of knowledge which have been opened up to us by its use.

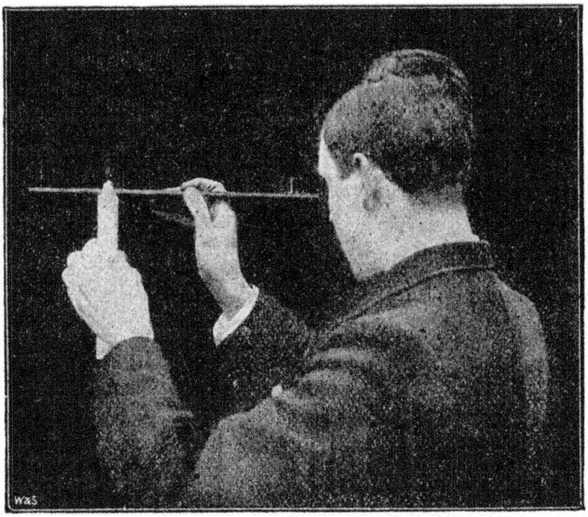

FIG. 3.—A simple form of spectroscope.

From an optician we can get a tiny prism for sixpence; glue it at one end of a piece of wood about 12 × 1 × ½ inch, so that we can see through it a coloured image of a needle stuck in at the other end of the piece of wood (Fig. 3). This we must do by looking sideways through it. Allow the needle to be illuminated by the flame of a spirit lamp into which salt is gradually allowed to fall. We see one image of the needle coloured in orange.

If we next illuminate the needle by a candle or a gas flame, taking care that the direct light from the candle does not fall upon the face of the prism, we then get no longer an image of the needle but a complete band of colour from red to blue. We have in fact an innumerable multitude of images of the needle close together.

If we finally go into the sunlight—taking care again to shield the prism, and allow a sunbeam to illuminate the needle, we get a spectrum of a different kind, inasmuch as it is full of dark lines, that is, some of the coloured rays are lacking and hence an image of the needle is not forthcoming. The positions of some of the chief dark lines lettered by Fraunhofer are shown in Fig, 11.

By such experiments as these, certain spectroscopic axioms have been formulated; three of them are very important.

First, when solid or liquid or densely gaseous bodies are incandescent, they give out continuous spectra.

Second, when a solid or liquid body reduced to a state of gas, or any gas itself, is giving light, the spectrum consists of bright images of the slit, and these are different for different substances.

Third, when light from a solid or liquid or densely gaseous body passes through the gas at a lower temperature, the gas absorbs those particular rays of light of which its own spectrum consists at that temperature.

We generally talk of "line" spectra for the reason that a narrow slit is employed, the image of which is a line. In these "lines" seen in the spectra of the heavenly bodies we have so many celestial hieroglyphics which we have to translate into chemical language by comparing their positions with those we observe in the spectra of terrestrial light sources. As for the determination of position we are enabled to do this with perfect definiteness, by considering the *wave-length* of the particular colour with which we have to deal. The wave-length is generally expressed by four figures, giving the length in ten millionths of a millimetre. Having these wave-lengths we may define the quality of every kind of light which reaches the human eye, whether from a terrestrial light source, the sun or any other celestial body.

By making careful wave-length comparisons between terrestrial and celestial spectra, we can determine, therefore, whether in different parts of space we have the same chemical substances, or substances perfectly and completely distinct; we can even go further, and say whether the substances are known or unknown to us here.

For the most powerful spectroscopes employed in modern research we are as much indebted to our American cousins as we are in the case of the largest refractors. Spectra are now for the most part obtained by the use of reflection from diffraction gratings. Rutherfurd was the first to silver a diffraction grating ruled on glass; Rowland followed him by ruling at once on speculum metal, and then made a step further in advance by using concave gratings, thus doing away with the use of lenses. All, or nearly all, the best recent work I shall have to appeal to in the succeeding chapters has been produced by means of

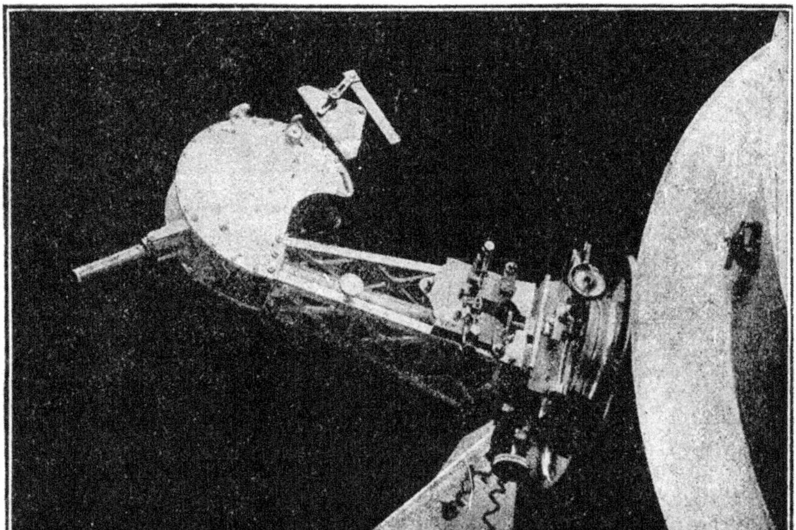

FIG. 4.—Star spectroscope, arranged for photography, attached to eye-end of reflecting telescope.

gratings, and one enormous advantage connected with this method is that the resultant spectrum is one of wave-lengths which are, therefore, more easily and accurately determined directly this way than indirectly by means of dispersion by prisms.

FIG. 5.—The eye-end of the Lick telescope with spectroscope attached.

For obtaining the spectra of the stars and nebulæ, one method is to attach a spectroscope of the ordinary kind at the eye-end of a telescope, whether refractor or reflector. And of course the more powerful and elaborate the instrument, the greater the accuracy in the determination of wave-length which is to be expected.

As an instance of one of these attachments I give a drawing of the eye-end of the Lick telescope with a spectroscope attached, for the importance of spectroscopic work was not lost sight of in the equipment of the Observatory, and a very powerful spectroscope is used in conjunction with the great equatorial for observing or photographing the spectra of the various celestial bodies.

The above, however, is not the only method of obtaining the spectra of stars, and indeed if photography be employed, the arrangement described in the next chapter does away with the employment of an ordinary spectroscope altogether.

CHAPTER III.—THE PRISMATIC CAMERA.

Description.

Fraunhofer, at the beginning of the present century found that to observe the spectra of stars the best thing to do was to put a prism outside the telescope, and to let the light enter the telescope and be brought to a focus after it had passed through the prism ; and it is a most unfortunate thing, that the neglect of the application of this principle has landed us probably in a delay of fifteen or twenty years in gathering knowledge on this subject. The whole credit of reviving this idea is due to Professor Pickering, of Harvard, who since its application has been able, with the aid of the large funds that he has at his disposal, and the magnificent help which he has accumulated round him, to obtain practically the spectra of all the stars down to the fifth or sixth magnitude in both hemispheres. In a few years' time we shall be able, thanks to his labours on the Draper Memorial, to work on the spectra of all the stars in both hemispheres, just as well as we can at present deal with their magnitudes and positions by the star charts.

My own attempts in this direction have had a much more limited scope. I have dealt with some of the brighter stars chiefly with a view to determining the sun's place among them.

The main instrument employed in this work has been a 6-inch refracting telescope, with an object-glass made and

FIG. 6.—Showing method of photographing stellar spectra by the objective prism.

corrected for G (see Fig. 11) by the Brothers Henry. This was used in conjunction with a prism of $7\frac{1}{2}°$ of dense glass by Hilger. The object-glass and prism are fixed at the end of a wooden tube, which is attached to the side of the 10-inch equatorial, at such an angle that the spectrum of a star falls on the middle of the photographic plate when its image is at the centre of the field of the larger instrument. The camera is arranged to take plates of the ordinary commercial size, $4\frac{1}{4} \times 3\frac{1}{4}$ inches. The spectra obtained with this instrument are 0·6 inch long from F to K. An excellent photograph of the spectrum of a first magnitude star can be obtained with an exposure of five minutes. Afterwards a 6-inch prism, with a refracting angle of 45°, obtained from the Brothers Henry, was used with the Henry 6-inch object glass. The spectra obtained with the latter are two inches long from F to K, and the definition is exquisite. In some photographs the calcium line at H is very clearly separated from the line of hydrogen, which occupies very nearly the same position. It is unnecessary to swing the back of the camera in order to get a perfect focus from F to K. The deviation of the prism is so great that it would be very inconvenient to incline the tube which supports it at the proper angle to the larger telescope. When photographing the spectrum of a star, therefore, the star is first brought to the centre of the field of the large telescope, and the proper deviation is then given by reading off on the declination circle. This method has been found to work quite satisfactorily.

With this combination the exposure required for a first magnitude star is about twenty minutes.

For the fainter stars, the 6-inch prism of $7\frac{1}{2}°$ has been adapted to a Dallmeyer rectilinear lens of 6 inches aperture and 48 inches focal length. At times two prisms of $7\frac{1}{2}°$ have been used on a 10-inch equatorial. The method of mounting the prisms is shown in Fig. 7.

The Clock Rate.

Since the spectrum of a point of light such as a star is a line so fine that the spectral lines would not be measurable, it is necessary to give it breadth. This is done by adjusting the prism so that the spectrum lies along a meridian of right ascension and altering the rate of the clock.

It is worth while to dwell a little on this clock error, and the way it is produced.

The proper regulation of this clock error and consequent "trail" of the spectrum across the plate parallel to itself are essential to the success of photographs taken by the objective prism. The spectrum of a bright star must obviously be made to trail more quickly than that of a fainter one, and a shorter exposure is sufficient. Since for the same clock error, and in the same time, a star near the pole will give a shorter trail than one nearer the equator, declination must also be taken into

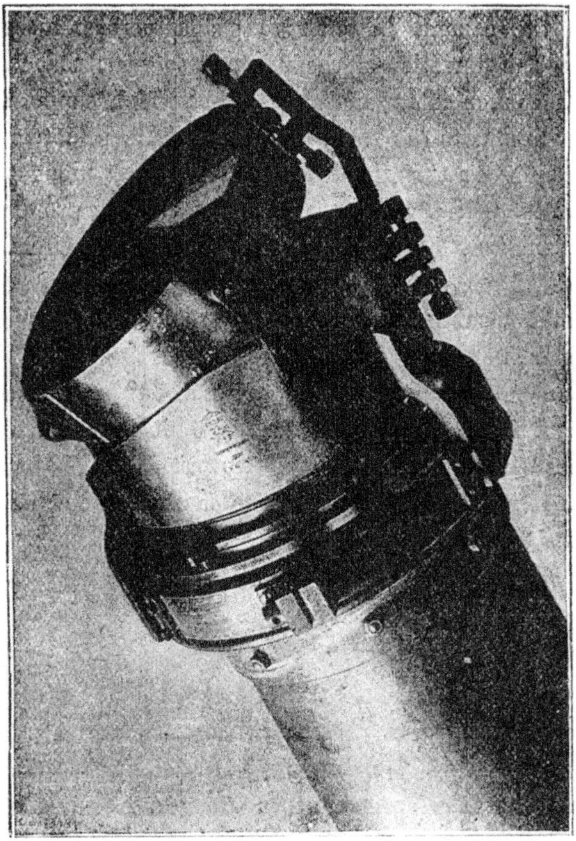

FIG. 7.—Objective prisms fitted to object-glass.

account. Keeping a constant clock error, equal widths of spectrum for stars of different declinations may be obtained by lengthening the time of exposure for stars away from the equator, but in that case the stars near the pole would be over-exposed in relation to those nearer the equator.

The exposure given to stars of equal magnitude should evidently be the same, no matter in what part of the sky they may be situated, and the clock error should therefore, be increased in proportion to the secant of the angle of declination.

The light-ratio of stars being $2{\cdot}512^n$, where n expresses the difference in magnitude, the time of exposure must vary in the same proportion, and the clock error in inverse proportion. Thus, where five minutes' exposure is sufficient for a first-magnitude star, thirty-one minutes is required to obtain a fully-exposed spectrum of a star of the third magnitude. This law, however, only applies to photographic magnitudes, and must be modified according to the type of spectrum or the colour of the star.

The red stars, being much weaker in blue and violet rays than the yellow or white stars, require much longer exposures than white stars of equal magnitude. To obtain a spectrum of β Pegasi extending to the K line, for example, at least three times the exposure required by a white star of similar magnitude must be given.

For conveniently adjusting the exposures, tables have been constructed which show at a glance the position of the regulator for a star of given magnitude and declination.

It is obvious that with an instrument of high dispersion, the number of stars it is possible to photograph is very limited, as the long exposures required for the fainter stars are impracticable, and, even if possible, the definition of the lines would be destroyed by atmospheric tremors.

Hence, it is at present only possible to photograph the spectra of the faint stars on a very small scale. With an objective of 8 inches aperture and 44 inches focal length, and a prism of 13° refracting angle, Professor Pickering has photographed the spectra of stars down to the eighth magnitude. These spectra are about 1 centimetre long, and a millimetre broad, and though they do not show a very great amount of detail, they are sufficient to reveal the type of spectrum.

With an instrument capable of photographing faint stars, a

large number of spectra may be taken at one exposure; but with the instruments of larger dispersion, this is not generally the case, as there are few bright stars of nearly equal magnitude sufficiently close together.

THE ELECTRICAL CONTROL.

In consequence of the great accuracy required in the driving of the telescope when long exposures are necessary, the 10-inch equatorial has been fitted with a simple and inexpensive form of electrical control. This is a modification of that designed by Mr. Russell, of the Sydney Observatory.[1] The existing driving gear has been altered so that the driving rod performs its revolution in a second, and the motion is then communicated to the driving screw through a small worm wheel. The driving rod is vertical and in two parts, the lower portion ending in a faced ratchet wheel, 3 inches in diameter, and with 200 teeth. The upper part of the rod ends in an arm at right angles to itself, and this arm carries a ratchet of suitable shape, held down by an adjustable spring. An electro-magnet connected with the controlling pendulum, is arranged so as to only permit the ratchet to pass it once a second (see Fig. 8). If the clock be driving too quickly, the ratchet is held until the stop is raised by the pendulum. When held in this way the ratchet is lifted out of the teeth, and the driving clock itself is not affected.

In order that this form of control may be effective, it is essential that the clock should be going a little too quickly, as the control is only capable of *retarding* the driving-rod.

The controlling pendulum is, of course, regulated to the rate required for the particular star which is being photographed.

In Mr. Russell's form of control the two parts of the driving rod are connected by friction plates. It was found, however, on testing this arrangement, that when the upper portion was

---

[1] *Monthly Notices*, vol. li, p. 43.

FIG. 8.—Electrical control for 10-inch equatorial.

held by the electro-magnet, the rate of the governors was seriously retarded; hence I introduced a ratchet wheel, and its working leaves nothing to be desired.

ENLARGEMENTS OF THE NEGATIVES.

Many of the negatives taken have been enlarged about nine times on glass, and further copies have been taken on bromide paper, bringing the enlargement up to about twenty-five times the size of the original.

Owing to various causes the photographic spectra obtained by the method of trails show irregularities resembling the lines along the spectrum observed when the slit of a spectroscope is partly clogged with dust. It has been noticed that the period of the irregularities is equal to the time of revolution of the main driving screw of the telescope, and hence they may be accounted for by supposing the driving gear to be mechanically imperfect.

In that case some of the parallel lines which by their juxtaposition form the broadened spectrum, are superposed, while others are drawn apart, thus giving rise to dark and bright lines parallel to the length of the spectrum. These lines are more apparent in the case of bright stars than fainter ones. If the telescope were driven with perfect regularity and the atmosphere were quite steady, we should obtain a spectrum of uniform intensity along its width. This condition has very nearly been obtained in some cases.

The irregularities above described are eliminated in the enlarged negatives by giving them a very slight up-and-down motion during exposure in a direction parallel to the lines of the spectrum. This was originally done by hand, but a negative holder has been constructed in which the necessary motion is given to the negative by a small driving clock.

A diagram of the arrangement is given below. The only

Fig. 9.—Negative holder used in enlarging.

drawback to this method is that defects of the film are apt to produce, by a succession of their images on the enlarging plate, lines (generally very faint) which have a semblance of the true spectrum lines.

To distinguish the real lines from the artificial ones, a direct enlargement of the spectrum is made on the same plate alongside the other, the to-and-fro motion being dispensed with. By a comparison of the two enlarged strips, one can see at a glance which are the true lines of the spectrum, and which are those produced by small irregularities on the film. It may be stated that Dr. Scheiner has also used a somewhat similar method to the one described, the only difference being that he caused the plate on which the enlargement was to be taken to have the oscillating motion, instead of the original negative. The method employed by me, though no account of it had been published, had been in use for some time before Dr. Scheiner's method was announced.[1]

The definition of the negatives obtained by means of the objective prism is of such excellence that they may be almost indefinitely enlarged, and this gives them a special value when we come to investigate the smaller differences between stars which have more or less resemblance to each other. Practically we are able to dispense with elaborate micrometric measurements, and by placing the enlargements alongside each other, to see at a glance which lines agree in position and which are different.

So much, then, for the recent vast improvement in our methods and instruments of research. It is, I think, abundantly clear that by means of the new aids which have been placed at our disposal, the recent improved condition of our stock-in-trade, and the wonderful diligence and skill of observers, chiefly in America, who have taken up the new work, we are now in a very much better position than we have ever been before to investigate all questions dealing with the chemistry and physics of the stars.

---

[1] *Nature*, vol. xlii, p. 303, 1890.

CHAPTER IV.—THE DISCOVERY OF HELIUM.

The Line D^3, 1868.

In the year 1868, spectrum analysis was first utilised in endeavouring to unravel the message which was conveyed to us by a most interesting eclipse observed in India. The diagrams will indicate the kind of record with which we have to deal in studying these celestial hieroglyphics. We are in one part dealing with the long waves of light, the red; we are in the other dealing with the shorter waves of light, the blue. The work done in that eclipse is indicated by the bright lines—the hieroglyphics—which, when translated as they have been, describe for us the chemical nature of the particular stuff in the sun, which made him put on a blood-red appearance "on his getting out of his eclipse." Taking the notes in the light scale which are lettered in the ordinary spectrum of sunlight, in order that they may be easily recognised and remembered, we learn the particular qualities of the light emitted by the blood-red streak.

We have one quality represented by the line D, another at C, and another at F. According to the diagram, one of the lines is in the position of D. One observer said it was "at D, or near D."

Soon after this eclipse was observed in India, a method, long before formulated, of studying the blood-red streak surrounding the sun without waiting for an eclipse was brought into operation.

By this method it was quite easy to make observations whenever the sun was shining, perfectly free from any of the difficulties attending the hurry and the worry and the excitement of an eclipse, which lasts only a few seconds.

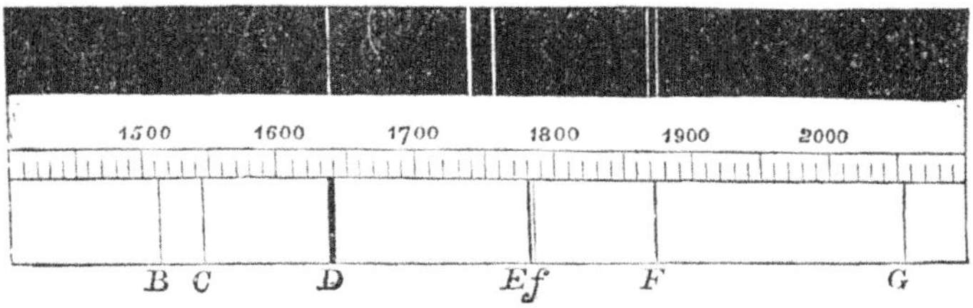

FIG. 10.—Pogson's diagram of the spectra of the sun's surroundings in the Eclipse of 1868. The bright lines seen are shown in the upper part of the diagram; the chief lines in the solar spectrum, red to the left, blue to the right, are shown in the lower part.

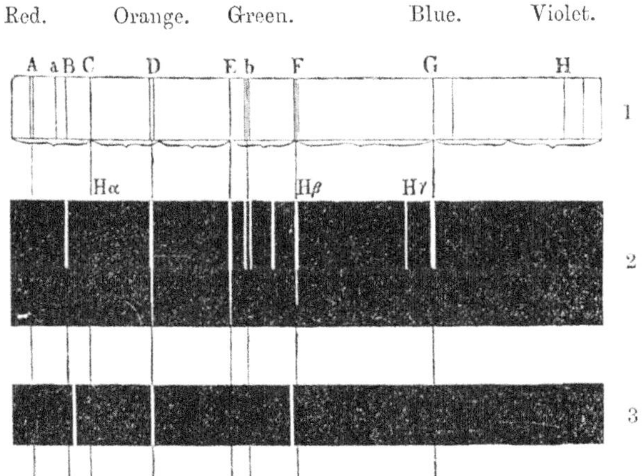

FIG. 11.—Summation of the observations of the spectrum of the sun's surroundings in the Eclipse of 1868. (1) Solar spectrum showing the position of the chief lines. (2) Rayet's observations of bright lines. (3) Herschel's observations of bright lines. (4) Tennant's.

Further, as the method consists of throwing an image of the sun, formed by a telescope, on to the slit of a spectroscope, so that the spectrum of the sun's edge and of the sun's surroundings can be seen at the same time, exact coincidence or want of coincidence between the bright and dark lines can be at once determined. During an eclipse this of course is not possible, as the ordinary spectrum of the sun, with its tell-tale dark lines, is invisible, because the sun, as we ordinarily see it, is hidden by the moon.

Working, then, under such very favourable conditions, it was seen that there was certainly a red line given by this lower part

C

FIG. 12.—The exact coincidence of the red line with the dark line C.

of the solar atmosphere coincident with the very important line in the solar spectrum which we call C. (Fig. 12.)

Another part of the spectrum in the blue-green was examined, and there again it was seen that the parts outside the sun gave us a bright line exactly in the position of the obvious dark line in the solar spectrum which is called F (Fig. 13); so that with regard to those two important lines, there was no doubt whatever that we were dealing with the substance which produces these dark lines in the solar spectrum.

Fig. 14 is a diagram of the yellow, or rather the orange, part of the solar spectrum, showing two very important lines, which are

FIG. 13.—The exact coincidence of the blue-green line with the dark line F.

called the lines D, due to the metal sodium, the investigation of which was just as important in solving the celestial hieroglyphics we call spectral lines as the Rosetta stone was important in settling the question of the Egyptian ones.

Pogson, in referring to the eclipse of 1868, said that the orange line was "at D, or near D." We see from the diagram,

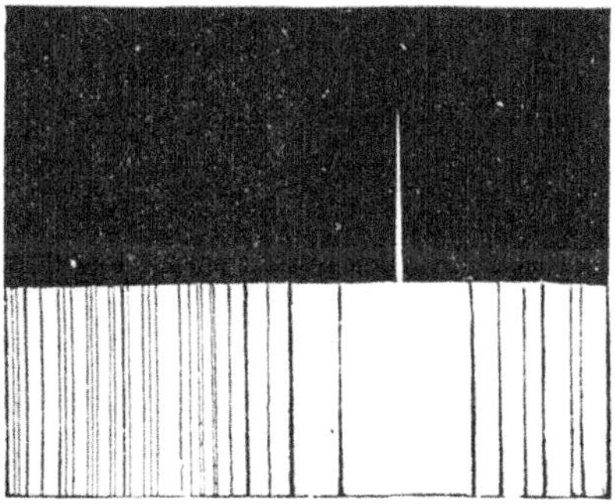

FIG. 14.—The want of coincidence of the orange line, D^3, with the dark lines D^1 and D^2.

Fig. 14, that the new method indicated that "near D" was the true definition. The line in this position in the spectrum, unlike the other two lines which I have indicated, has no connection at all with any of the dark lines in the ordinary solar spectrum. We were therefore perfectly justified in attaching considerable importance to this divergence in the behaviour of this line, taking the normal behaviour to be represented by the two strong lines in the red and the blue-green. The new line was called D^3 to distinguish it from the sodium lines D^1 and D^2.

A considerable amount of work was done with regard to the orange line. It was found that there was no substance in our laboratories which could produce it for us, whereas in the case of the line D we simply had to burn some sodium, or even common salt, in a flame to produce it; and the other lines in the red and the blue-green were easily made manifest by enclosing hydrogen in a vacuum tube and passing an electric current through it, or observing the spectrum of a spark in a stream of coal-gas.

Now at the first blush it looked very much as if this line was really due to the same element which produced the others at C and F, and it was imagined that the reason we did not see it in our laboratories was because it was a line which required a very considerable thickness of hydrogen to render it visible. That was the first idea, and Dr. Frankland and myself found that there was very considerable justification for this view, because a simple calculation showed that the thickness of the solar atmosphere, which was producing that orange line under the conditions which enabled us to see it in our instruments by looking along the edge of the sun, was something like 200,000 miles.

Hence, in order to get a final decision on this point, there was nothing for it but to tackle the question from a perfectly different point of view, and the different point of view was this. The work had not gone on very long before one found minute alterations in the positions of these lines in the spectrum; the orange line, for instance, might sometimes be slightly on one side, and sometimes on the other, of its normal position. Further work showed that in these so-called "changes of wave-length" we had a precious means of determining the rate of movement of the gases and vapours in the solar atmosphere.

Fig. 15 indicates how these changes of wave-lengths are shown in the spectroscope. The lines are contorted in both directions, and sometimes to a very considerable extent, indicating wind movements on the sun, reaching, and sometimes exceeding, 100 miles *a second.*

Blue. Red.

A B C

FIG. 15.—Changes of wave-length of the F hydrogen line when a solar cyclone is observed. A, the change towards the blue indicates the advancing side of cyclone. C, the change towards the red indicates the retreating side. B, the whole cyclone is included in the width of the slit, and both changes of wave-length are visible.

We had here a means of determining whether the orange line was produced by the same gases which gave the red and blue lines, because if so, when we got any alteration in the position of the red and blue lines, which always worked together, we should get an equivalent alteration in the position of the orange one.

I found that the orange line behaved quite differently from either the red or the blue lines; so then we knew that we were not dealing with hydrogen; hence we had to do with an element which we could not get in our laboratories, and therefore I took upon myself the responsibility of coining the word helium, in the first instance for laboratory use. At the time I gave the name I did not know whether the substance which gave us the line D^3 was a metal like calcium or a gas like hydrogen, but I did know that it behaved like hydrogen and that hydrogen, as Dumas had stated, behaved as a metal.

This kind of work went on for a considerable time, and what one found was, that very often in solar disturbances we certainly were dealing with some of the lines of substances with which we are familiar on this earth; but at the same time it was very remarkable that when the records came to be examined, as they ultimately were with infinite care and skill, it was found that not only did we get this line in the orange indicating an unknown element associated with substances very well known, like magnesium, but that there were many other unknown lines as well. Within a few months of my first observations, several new lines about which nothing was known were thus observed.

The Discovery of other Unknown Lines, 1869.

The place of the orange line D^3 I determined on 20th October 1868. Among many other lines behaving like it, two at wavelengths 4923 and 5017 were discovered in June, 1869, and afterwards another at 6677, while Professor Young noted another in September, 1869, at 4471. He wrote:—

"I desire to call special attention to 2581·5 on Kirchhoff's scale, the only one of my list, by the way, which is not given on Mr. Lockyer's. This line, which was conspicuous at the Eclipse of 1869, seems to be *always present* in the spectrum of the chromosphere. . . . It has no corresponding dark line in the ordinary solar spectrum, and not improbably may be due to the same substance that produces D^3."

The wave-length of this line on Rowland's scale is 4472.

This same line was noted also by Lorenzoni, and named *f*. Another line at 4026 was added later by Professor Young.

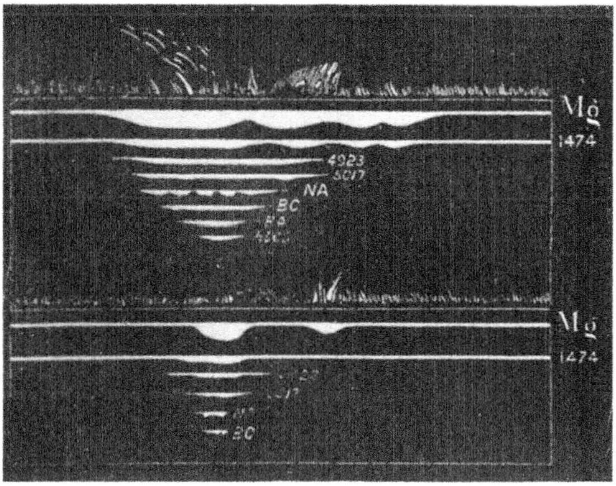

FIG. 16.—Tacchini's observations of two slight solar disturbances showing the height to which the layers of the different gases extend. Magnesium vapour is highest of all, and is furthest extended; next comes a gas of still unknown origin, indicated by a line at 1474 of Kirchhoff's scale, and so on.

Then with regard to solar disturbances. Let me refer in detail to a diagram indicating some results arrived at by the Italian observers (Fig. 16). We are dealing with the spectroscopic record of two slight disturbances in a particular part of the sun's atmosphere. The spectroscope tells us that in that region there was a quantity of the vapour of magnesium which is collected in that place. Then we find that another substance, about which we again know nothing whatever, is also visible in that region, and then we get the further fact that in those particular disturbances we get four other spectral lines indicated as being disturbed, and of those four lines we only know about one.

In that way it very soon became perfectly clear to those who were working at the sun, that in all these disturbances, or at all events in most of them, we were dealing to a large extent with lines not seen in our laboratories when dealing with terrestrial substances; this work went on till ultimately, thanks to the labours of Professor Young in America, we had a considerable list of lines coming from known and unknown substances which had been observed under these conditions in solar disturbances, and Professor Young was enabled to indicate the relative number of times these lines were visible. For instance, the lines which are most frequently seen under these conditions he tabulated as represented by the number 100, and of course the line which was least frequently seen would be represented by 1; and therefore from these so-called "frequencies" we got a good idea of the number of times we might expect to see any of these disturbance-lines when anything was going on in the sun.

It was this kind of work which made Tennyson write those very beautiful lines:

"Science reaches forth her arms
To feel from world to world."[1]

Dr. Hillebrand's Researches on Uraninite, 1888.

In this year Dr. Hillebrand, one of the officials in the Geological Department at Washington, was engaged upon the chemical examination of specimens of the mineral uraninite from various localities.

He dealt with crystals which he put in a vessel containing

---

[1] And then he added:

" and charms
Her secret from the latest moon."

I mention this because Tennyson, whose mind was saturated with astronomy, had already grasped the fact that what had already been done was a small matter compared with what the spectroscope could do; and now the prophecy is already fulfilled, for by means of the spectroscopic examination of the light from the stars we can tell that some of them are double stars, that is to say, in poetic language, stars with attendant moons. Although we can thus charm the secret from each moon by means of the spectroscope, to see the moon it would require (in the case of β Aurigæ) a telescope not 80 feet long, but with an object-glass 80 feet in diameter, because the closer two stars are together the greater must be the diameter of the object-glass, independently of its focal-length and magnifying power.

some sulphuric acid and water. He found that bubbles of gas were produced out of the crystal by means of the sulphuric acid. He collected this gas and came to the conclusion that it was nitrogen.

This result was new. He thus wrote about it:—

"In consequence of a certain observation" [the one I have just referred to] "and its results, an entirely new direction was given to the work, and its scope wonderfully broadened. This was the discovery of a hitherto unsuspected element in uraninite, existing in a form of combination not before observed in the mineral world."

It is not needful here to follow Dr. Hillebrand through all the painstaking and patient labour he cut out for himself to explain this anomalous behaviour. Needless to say he did not omit to employ the spectroscope to test the nature of the new gas.

His observations were thus described:—[1]

"In a Geissler tube under a pressure of 10 millimetres and less, the gas afforded the fluted spectrum of pure nitrogen as brilliantly and as completely as was done by a purchased nitrogen tube. In order that no possibility of error might exist, the tube was then reopened and repeatedly filled with hydrogen, and evacuated till only the hydrogen lines were visible. When now filled with the gas and again evacuated, the nitrogen spectrum appeared as brilliantly as before, with the three bright hydrogen lines added."

On this paragraph I may remark that it has long been known that gases like nitrogen give us quite distinct spectra at different temperatures—one fluted, another containing lines. Which of these we shall see in a tube will depend upon the pressure of the gas, and the electric current used. The fluted spectrum of nitrogen is very bright and full of beautiful detail in the orange part of the spectrum; the line spectrum, on the other hand, is almost bare in that region.

It is important to note *that it so happened* that the pressure

---

[1] "On the Occurrence of Nitrogen in Uraninite," *Bulletin*, No. 78, U.S. Geol. Survey, 1889–90, p. 55.

and electric conditions employed by Dr. Hillebrand enabled him generally to see the fluted spectrum. This, however, was not always the case. In an interesting letter to Professor Ramsay he writes (*Proc. Roy. Soc.*, vol. lviii, p. 81) :—

"Both Dr. Hallock and I observed numerous bright lines on one or two occasions, some of which apparently could be accounted for by known elements—as mercury, or sulphur from sulphuric acid; but there were others which I could not identify with any mapped lines. The well-known variability in the spectra of some substances under varying conditions of current and degree of evacuation of the tube led me to ascribe similar causes for these anomalous appearances, and to reject the suggestion made by one of us in a doubtfully serious spirit, that a new element might be in question."

Dr. Hillebrand concludes his paper as follows :—

"The interest in the matter is not confined merely to a solution of the composition of this one mineral; it is broader than that, and the question arises: May not nitrogen be a constituent of other species in a form hitherto unsuspected and unrecognisable by our ordinary chemical manipulations? And, if so, other problems are suggested which it is not now in order to discuss."

I shall show in the sequel that if Dr. Hillebrand had gone a little further in his researches he would have discovered the line D^3 and other lines to which reference has already been made in the gases he had obtained from uraninite.

D^3 and other Unknown Lines in Nebulæ, 1890.

A negative of the nebula of Orion, taken at my observatory at Westgate-on-Sea in 1890 contains fifty-six lines, and of course by determining, as we have been able to do approximately, the wave-lengths—the positions of these lines in the spectrum—we can determine the exact light notes represented, and therefore the substances which produce them. In this spectrum of the nebula of the Orion were lines of unknown origin exactly coinciding with those unknown lines which I have already referred to as having been seen in the sun's atmosphere. Some of the unknown lines in that atmosphere, those that we have not been able to see in our laboratories, are identical in position with some of the unknown lines in the nebula of Orion.

I may remark that as early as 1886 Dr. Copeland had discovered D^3 in the visible spectrum of the nebula, and in a letter to him I had suggested that another line he had recorded at 447 might be Lorenzoni's *f*; this he thought to be probable. The matter was set for ever at rest by the photograph which established the presence of 4471 and 4026, already noted as a solar line as well.

Professor Campbell, of the Lick Observatory, obtained other photographs of the spectrum of the nebula some two or three years after mine was taken. In the following list of lines in my photograph an asterisk denotes that Campbell gives a line nearly in the same position. He recorded no line which did not appear on my photograph—

3896*
3888*
4011
4026*
4121*
4143*
4168
4390*
4472*
4716*
4924
5875·8 = D^3

The same Unknown Lines occur in the Stars, 1892.

About the year 1890 I began the photography of stellar spectra at Kensington, with special reference to their classification on the basis of the chemical constituents established by their spectra. By 1892 several important results had been obtained, while the progress of this branch of science lately has been so considerable that any statement regarding the positions of lines, and, therefore, the chemical origins of them may be made with a considerable amount of certainty as depending upon very accurate work.

The various classes in which the stars have been classified by different observers, according to their spectra, are discussed elsewhere, but some of the more salient differences must be pointed out here; thus we have stars with many lines in their spectra, others with comparatively few. I will take the many-lined stars first.

The diagram (Fig. 17) represents the spectrum of Arcturus, a star the spectrum of which closely resembles that of the sun. In α Cygni we have another star with many lines; but here we note, when we leave the hydrogen on one side and deal with the other stronger lines, that there is little relation between the solar spectrum and these lines.

I next come to the stars with few lines: these are well represented by many of the chief stars in the constellation of Orion. Bellatrix is given as an example (Fig. 18).

Then I have next to state, that in the photographs of the spectra of many stars, chiefly of those more or less like Bellatrix, we found the same lines which we have, so far, classified as unknown, for the reason that in our laboratories we have not been able to get any lines which correspond with them. I again mention D^3, 4471 and 4026, previously noted as appearing both in the chromosphere and in the nebula of Orion.

But the thing is much more interesting even than this; not only these but *all* the chief unknown lines appearing in the nebula of Orion are also found in these stars. And this is so absolutely true that there is no necessity to give a list of the unknown lines seen in Bellatrix; every one of them given in the nebula has found its place, and, so far, practically, no others.

This, of course, marked a great development of the inquiry, and made the question of the unknown lines more important than ever.

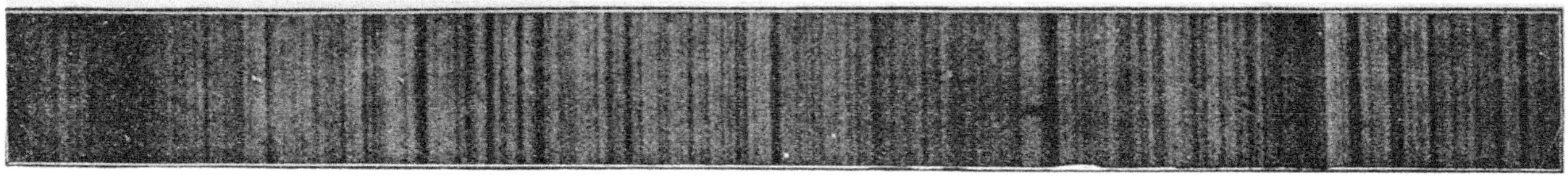

FIG. 17.—Spectrum of Arcturus between G and K.

FIG. 18.—The spectrum of Bellatrix between H and F.

FIG. 19.—Untouched reproduction of photograph (African station) taken very shortly after the commencement of totality, the exposure being "instantaneous." At this phase of the eclipse a considerable arc of the chromosphere was visible, and its spectrum is therefore shown in addition to the spectrum of the higher reaches of some of the large prominences extending beyond the moon's limb. It will be seen that at H and K there are long arcs of chromosphere and prominences, the absent portions being of course obscured by the moon. One very small prominence is especially rich in lines.

PHOTOGRAPHIC RESULTS DURING A SOLAR ECLIPSE, 1893.

A method which was first employed by Respighi and myself during the eclipse of 1871, was used on a large scale and with great effect during the eclipse of 1893. The light proceeding from the luminous ring round the dark moon was made to give us a series of rings, representing each bright line seen by the ordinary method, on a photographic plate. The observers this time were stationed in West Africa and in Brazil. The African station was up one of the rivers, not very far away from the town of Bathurst. The Brazilian station was near Para Curu. The same instrument which was previously referred to as used for obtaining photographs of the stars was sent to the African station in order that photographs of the eclipse of the sun might be taken on exactly the same scale as the photographs of the stars had been, so that the stellar and solar records in the photographs might be compared. The results obtained by Messrs. Fowler and Shackleton, who were in charge of the instruments at the two stations, will be gathered from the accompanying diagrams, Figs. 19 and 20.

We get more or less complete rings when we are dealing with an extended arc of the chromosphere, or lines of dots when any small part of it is being subjected to a disturbance which increases the temperature and, possibly, the numbers of the different vapours present.

The efficiency of this method of work, with the dispersion employed, turns out to be simply marvellous, and, in securing such valuable and permanent records as these, we have done very much better than if we had contented ourselves with the style of observations that I have referred to as having been made in 1868 (see p. 29).

As was expected, the comparison between solar and stellar records thus rendered possible, enabled a very great advance to be made.

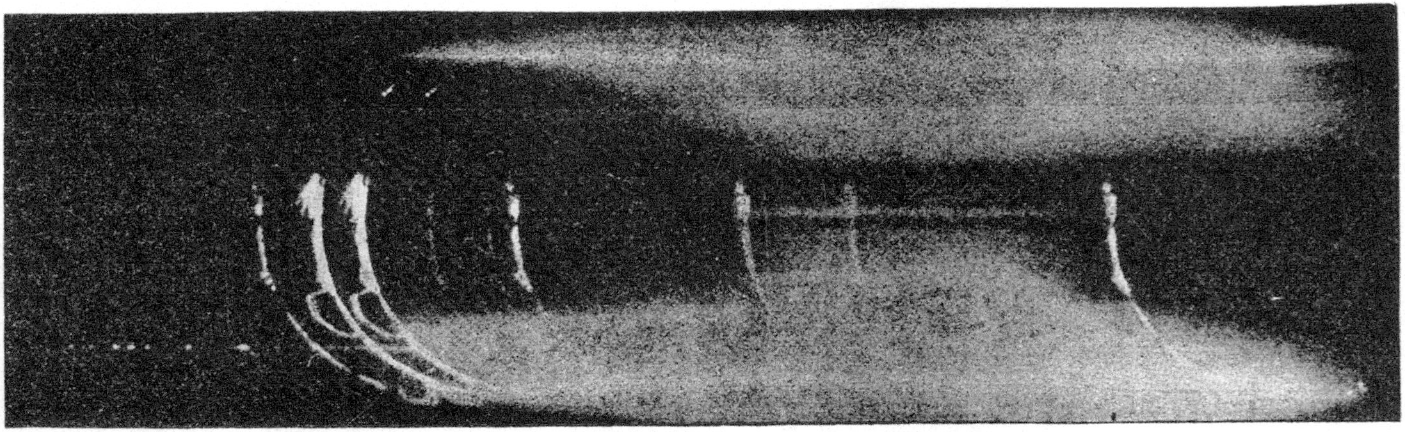

Fig. 20.—Photograph 21 (African station) taken shortly before the end of totality. A portion of the chromosphere on the other edge of the dark moon is now visible in addition to numerous prominences. It will be seen that one of the smallest prominences is rich in lines, and closely resembles that which appears in Fig. 19.

On examining these eclipse records, we find that we have to do exactly with those unknown lines which had already been photographed in the stars and in the nebulæ.

As was to be expected, we, of course,. deal with the lines recorded in the first observations of the solar disturbances, and chronicled in that table of Professor Young's to which I have already called attention; but the important thing is the marvellously close connection between eclipse- and star-spectrum photographs so far as the "unknown lines" are concerned.

Nearly all the lines given in the table on p. 39 as visible in the nebula of Orion, and afterwards found in Bellatrix, are among the lines photographed during the eclipse.

CHAPTER V.—DISCOVERY OF A TERRESTRIAL SOURCE OF HELIUM, 1895.

THE NEW MINERAL GASES.

The year 1894 was made memorable by the announcement of the discovery by Lord Rayleigh and Professor Ramsay of a new gas called argon, and it is known that the discovery was brought about chiefly, in the first instance, by the very accurate observations of Lord Rayleigh, who found that when he was determining the weight of nitrogen in a globe of a certain capacity, the weight depended upon the source from which he obtained the gas.

From the nitrogen from atmospheric air he obtained one weight, and from that obtained by certain chemical processes he obtained another, and, ultimately, it was found that there was an unknown element which produced these results, these various changes in the weight, and, as a consequence, we had in 1895 the discovery of argon.

Early in 1895 it struck Mr. Miers, of the British Museum, that it might be desirable to draw attention to the nitrogen which we have seen Dr. Hillebrand, in 1888, obtaining from his crystal of uraninite; his observations, of course, were more in the mind of Mr. Miers than in the minds of the pure chemists. He, therefore, communicated with Professor Ramsay, who lost no time, because it was very interesting to study every possible source of nitrogen and see what its behaviour was in regard to the quantity of argon associated with it, and in the relation generally of the gas to the argon which was produced from it.

Professor Ramsay treated uraninite in exactly the same way that Dr. Hillebrand had done in 1888. The gas obtained as Dr. Hillebrand had obtained it was eventually submitted to a spectroscopic test, following Dr. Hillebrand's example. But here a noteworthy thing comes in.

It so happened that the pressure and electrical conditions employed by Professor Ramsay were so different from those used by Dr. Hillebrand that, although nitrogen was undoubtedly present, the fluted spectrum which, as I have previously stated, floods the orange part of the spectrum with luminous details, was absent. But still there was *something* there.

Judge of Professor Ramsay's surprise when he found that he got a bright orange line; that was the chief thing, and *not* the strong suggestion of the spectrum of nitrogen. Careful measurements indicated that the twenty-six-year-old helium had at last been run to earth, D^3 was at last visible in a laboratory. Professor Ramsay was good enough to send specimens of the tubes containing this gas round to other people, and he sent one of them to me.

I received Professor Ramsay's tube on 28th March, but it was not suitable for the experiments I wished to make.

On 29th March, therefore, as Professor Ramsay was absent from England, in order not to lose time I determined to see whether the gas which had been obtained by chemical processes would not come over by heating *in vacuo*, after the manner described by me to the Royal Society in 1879,[1] and Mr. L. Fletcher was kind enough to give me some particles of uraninite (bröggerite) to enable me to make the experiment.

This I did on 30th March, and it succeeded; the gas giving

[1] *Proc. Roy. Soc.*, vol. xxix, p. 266.

the orange line came over, associated with hydrogen, in good quantity.

From 30th March onwards my assistants and myself had a very exciting time. One by one the unknown lines I had observed in the sun in 1868 were found to belong to the gas I was distilling from bröggerite. Not only D^3 but 4923, 5017, 4471 (Lorenzoni's *f*), 6677 (the BC of Fig. 16), referred to previously, and many other solar lines, were all caught in a few weeks.

But this was by no means all. The solar observations had been made by eye, and referred therefore to the less refrangible part of the spectrum, but I had obtained hundreds of stellar photographs, so I at once proceeded to photograph the spectrum of the gas and compare its more refrangible lines with stellar lines.

Here, if possible, the result was still more marvellous. In the few-lined stars by 6th May I had caught nearly all the most important lines at the first casts of the spectroscopic net. Fig. 24, which includes some later results, will give an idea of the tremendous revelation which had been made as to the chemistry of some of the stages of star-life. I pointed out on 8th May that we had already "run home" the most important lines in the spectra of a certain group of stars, in which alone we find D^3 reversed.

These results enabled us at once to understand how it was that the "unknown lines" had been seen both in the sun's chromosphere and some nebulæ and stars. The gas obtained from the minerals made its appearance in various heavenly bodies, and because it was a gas it made its appearance in different temperature conditions; and the more the work goes on, we find that this gas is really the origin of many, but certainly not of *all*, of the unknown lines which have been teasing astronomical workers for the last quarter of a century.

The First Investigations of the Spectrum of the Gas from Cleveite.

The dates of the papers communicated to the Royal Society recording the observations of the lines in the gas obtained from minerals which had been previously recorded are as follows:—

25th April	4471	4144	
8th May	667	4388	4026
9th May	3889		
28th May	7065		
29th May	5048	5016	4922

The lines at 667 and 5016 had been previously seen by Thalén (*Comptes Rendus,* 16th April, 1895).

Although the general distribution and intensities of the lines in the gases from bröggerite and cleveite sufficiently corresponded with some of the chief "unknown lines" in the solar chromosphere and some of the stars to render identity probable, it was necessary to see how far the conclusion was sustained by detailed investigations of the wave-lengths of the various lines.

Employment of High Dispersion.

This was practically a separate branch of the work, as the observations had to be made in the observatory. Next I give here the observations relating to D^3 and 4471.

The Orange Line, λ 5875·9.—Immediately on receiving from Professor Ramsay, on 28th March, a small bulb of the gas obtained from cleveite, a provisional determination of wave-length was made by Mr. Fowler and myself, in the absence of the sun, by micrometric comparisons with the D lines of sodium, the resulting wave-length being 5876·07 on Rowland's scale. It was at once apparent, therefore, that the gas-line was not far removed from the chromospheric D^3, the wave-length of which is given by Rowland as 5875·98.

The bulb being too much blackened by sparking to give sufficient luminosity for further measurements, I set about preparing some of the gas for myself by heating bröggerite *in vacuo*, in the manner I have already described. A new measurement was thus secured on 30th March, with a spectroscope having a dense Jena glass prism of 60°; this gave the wave-length 5876·0.

On April 5th I attempted to make a direct comparison with the chromospheric line, but though the lines were shown to be excessively near to each other, the observations were not regarded as final.

Professor Ramsay having been kind enough to furnish me, on 1st May, with a vacuum tube which showed the orange line very brilliantly, a further comparison with the chromosphere was made on 4th May. The observations were made by Mr. Fowler, in the third order spectrum of a grating having 14,438 lines to the inch, and the observing telescope was fitted with a high power micrometer eye-piece; the dispersion was sufficient to easily show the difference of position of the D^3 line on the east and west limbs, due to the sun's rotation. Observations of the chromosphere were therefore confined to the sun's poles.

During the short time that the tube retained its great brilliancy, a faint line, a little less refrangible than the bright orange one, and making a close double with it, was readily seen; but afterwards a sudden change took place, and the lines almost faded away. While the gas-line was brilliant, it was found to be "the least trace more refrangible than D^3, about the thickness of the line itself, which was but narrow" ("Observatory Note Book"). The sudden diminution in the brightness of the lines made subsequent observations less certain, but the instrumental conditions being slightly varied, it was thought that the gas-line was probably less refrangible than the D^3 line by about the same amount that the first observation showed it to be more refrangible. Giving the observations equal weight, the gas-line would thus appear to be probably coincident with the middle of the chromospheric line, but if extra weight were given to the first observation, made under much more favourable conditions, the gas-line would be slightly more refrangible than the middle of the chromosphere line.

Pressure of other work did not permit the continuation of the comparisons. In the meantime, Runge and Paschen announced (*Nature*, vol. lii, p. 128) that they also had seen the orange line of the cleveite gas to be a close double, neither component having exactly the same wave-length as D^3, according to Rowland.

They give the wave-length of the brightest component as 5878·883, and the distance apart of the lines as 0·323.

This independent confirmation of the duplicity of the gas-line led me to carefully re-observe the D^3 line in the chromosphere for evidences of doubling. On 14th June observations were made by Mr. Shackleton and myself of the D^3 line in the third and fourth order spectra under favourable conditions.

"The line was seen best in the fourth order, on an extension of the chromosphere or prominence on the north-east limb of the sun. The D^3 line was seen very well, having every appearance of being double, with a faint component on the red side, dimming away gradually; the line of demarcation between the components was not well marked, but it was seen better in the prominence than anywhere else on the limb." ("Observatory Note Book.")

It became clear, then, that the middle of the chromosphere line, as ordinarily seen, and as taken in the comparison of 4th May, does not represent the place of the brightest component of the double line, so that exact coincidence was not to be expected.

The circumstance that the line is double in both gas and chromosphere spectrum, in each the less refrangible component being the fainter, taken in conjunction with the direct comparisons which have been made, rendered it highly probable that one of the gases obtained from cleveite is identical with that which produces the D^3 line in the spectrum of the chromosphere.

Other observers have since succeeded in resolving the chromospheric line. On 20th June Professor Hale found the line to be clearly double in the spectrum of a prominence, the less refrangible component being the fainter, and the distance apart of the lines being measured as 0·357 tenth-metres (*Ast. Nach.*, 3302). The doubling was noted with much less distinctness in the spectrum of the chromosphere itself on 24th June. Professor Hale points out that Rowland's value of the wave-length (as well as that of 5875·924, determined by himself on 19th and 20th June) does not take account of the fact that the line is a close double.

Dr. Huggins, after some failures, observed the D^3 line to be double on 10th July (*Ast. Nach.*, 3302); he also notes that the less refrangible component was the fainter, and that the distance apart of the lines was about the same as that of the lines in the gas from cleveite, according to Runge and Paschen.

It may be added that, in addition to appearing in the chromosphere, the D^3 line has been observed as a bright line in nebulæ by Dr. Copeland, Professor Keeler, and others; in β Lyræ and other bright line stars; and as a dark line in such stars as Bellatrix, by Mr. Fowler, Professor Campbell, and Professor Keeler. In all these cases it is associated with other lines, which, as I shall show presently, are associated with it in the spectra of the new gases.

The Blue Line, λ 4471·6.—A provisional determination on 2nd April of the wave-length of a bright blue line, seen in the spectrum of the gases obtained from a specimen of cleveite, showed that it approximated very closely to a chromospheric line, the frequency of which is stated as 100 by Young.

This line was also seen very brilliantly in the tube supplied to me by Professor Ramsay on 1st May, and on 6th May it was compared directly with the chromosphere line by Mr. Fowler.

The second order grating spectrum was employed. The observations in this region were not so easy as in the case of D^3, but with the dispersion employed, the gas-line was found to be coincident with the chromospheric one. In this case also the chromosphere was observed at the sun's poles, in order to eliminate the effects due to the sun's rotation.

Besides appearing in the spectrum of the chromosphere, the line in question is one of the first importance in the spectra of nebulæ, bright line stars, and of the white stars such as Bellatrix and Rigel.

The Infra-red Line, λ 7065·5.—In addition to D^3 and the line at 4471·6, there is a chromospheric line in the infra-red which also has a frequency of 100, according to Young. On 28th May I communicated a note to the Royal Society, stating that this line had been observed in the spectrum of the gases obtained from bröggerite and euxenite (*Proc. Roy. Soc.*, vol. lviii, p. 192), solar comparisons having convinced me that the wave-length of the gas-line corresponded with that given by Young; and I added:—

"It follows, therefore, that besides the hydrogen lines all three chromospheric lines in Young's list which have a frequency of 100 have now been recorded in the spectra of the new gas or gases obtained from minerals by the distillation method."

M. Deslandres, of the Paris Observatory, has also observed the line at 7065 in the gas obtained from cleveite (*Comptes rendus*, 17th June, 1895, p. 1331).

A great deal of work has been done upon these gases from other points of view than those which affect their cosmical relations, and perhaps I may be allowed next to refer to some of the results which have been obtained by myself.

The Cleveite Gases not connected with Argon.

The first point is that the gas from the minerals contains no argon. Dr. Ramsay in his first experiments came to the conclusion that the spectra of argon and helium contained many common lines; indeed, at first, the observed coincidences were so remarkable that he came to the conclusion that the connection was so close that atmospheric argon contained a gas absent from the argon seen in his helium tube.

This statement was subsequently withdrawn, but the compound nature both of argon and helium was suggested by the fact that there were lines common to the two gases. These lines were in the red; one coincidence I found broke down with moderate dispersion, the other yielded subsequently to the still greater dispersion employed by Drs. Runge and Paschen. It may be also stated here that I have not found a single coincidence between argon and any line in the spectrum of any celestial body whatever. This happens, as everybody knows, also in the case of oxygen, nitrogen, chlorine and the like.

THE CLEVEITE GAS A MIXTURE.

The first spectroscopic observations made it perfectly obvious that the gas as obtained from uraninite is a mixture of gases; that the gas which gives the yellow line is not an isolated one, but is mixed up with other gases which give other lines.

In May I wrote as follows:—[1]

"The preliminary reconnaissance suggests that the gas obtained from bröggerite by my method is one of complex origin.

"I now proceed to show that the same conclusion holds good for the gases obtained by Professors Ramsay and Clève from cleveite.

"For this purpose, as the final measures of the lines of the gas as obtained from cleveite by Professors Ramsay and Clève have not yet been published, I take those given by Crookes and Clève, as observed by Thalén.

"The most definite and striking result so far obtained is that in the spectra of the minerals giving the yellow line I have so far examined, I have never once seen the lines recorded by Crookes and Thalén in the blue. This demonstrates that the gas obtained from certain specimens of

[1] *Proc. Roy. Soc.*, vol. lviii, p. 114.

cleveite by chemical methods is vastly different from that obtained by my method from certain specimens of bröggerite, and since, from the point of view of the blue lines, the spectrum of the gas obtained from cleveite is more complex than that from bröggerite, the gas itself cannot be more simple.

"Even the blue lines themselves, instead of appearing *en bloc*, vary enormously in the sun, the appearances being

"4922 (4921·3) = thirty times
"4713 (4712·5) = twice.

"These are not the only facts which can be adduced to suggest that the gas from cleveite is as complex as that from bröggerite."

It is seen that quite early in the inquiry we had not only spectroscopic evidence in the laboratory which was complete in itself, but that the case was greatly strengthened when the behaviour of the various lines in the sun and stars was also brought into evidence.

In the first case we had the laboratory separation of D^3 from the lines 5048, 5016, and 4922.

Later on in the same month I showed that the lines at D^3 and 447 behaved in one way, and that at 667 behaved in another.

In order to test this view I made some observations based on the following considerations:—

(1) In a simple gas like hydrogen, when the tension of the electric current given by an induction coil is increased by inserting first a jar and then an air-break into the circuit, the effect is to increase the brilliancy and the breadth of all the lines, the brilliancy and breadth being greatest when the longest air-break is used.

(2) Contrariwise, when we are dealing with a known compound gas; at the lowest tension we may get the complete spectrum of the compound without any trace of its constituents, and we may then, by increasing the tension, gradually bring in the lines of the constituents, until, when complete dissociation is finally reached, the spectrum of the compound itself disappears.

Working on these lines the spectrum of the spark at atmospheric pressure passing through the gas or gases distilled from bröggerite, has been studied with reference to the special lines C (hydrogen), D^3, 667, and 447.

The first result is that all the lines do not vary equally as they should do if we were dealing with a simple gas.

The second result is that at the lowest tension 667 is relatively more brilliant than the other lines; on increasing the tension, C and D^3 considerably increase their brilliancy, 667 relatively and absolutely becoming more feeble, while 447, seen easily as a narrow line at low tension, is almost broadened out into invisibility as the tension is increased in some of the tubes, or is greatly brightened as well as broadened in others (Fig. 21).

FIG. 21.—Diagram showing changes in intensities of lines brought about by varying the tension of the spark. 1. Without air-break. 2. With air-break.

The above observations were made with a battery of five Grove cells; the reduction of cells from five to two made no difference in the phenomena except in reducing their brilliancy.

Reasoning from the above observations, it seems evident that the effect of the higher tension is to break up a compound or compounds, of which C, D^3, and 447 represent constituent elements; while, at the same time, it would appear that 667 represents a line of some compound which is simultaneously dissociated.

The unequal behaviour of the lines has been further noted in another experiment, in which the products of distillation of bröggerite were observed in a vacuum tube and photographed at various stages. After the first heating D^3 and 4471 were seen bright, before any lines other than those of carbon and hydrogen made their appearance. With continued heating 667, 5016, and 492 also appeared, although there was no notable increase of brightness in the yellow line; still further heating introduced additional lines 5048 and 6347.

These changes are represented graphically in the following diagram (Fig. 22).

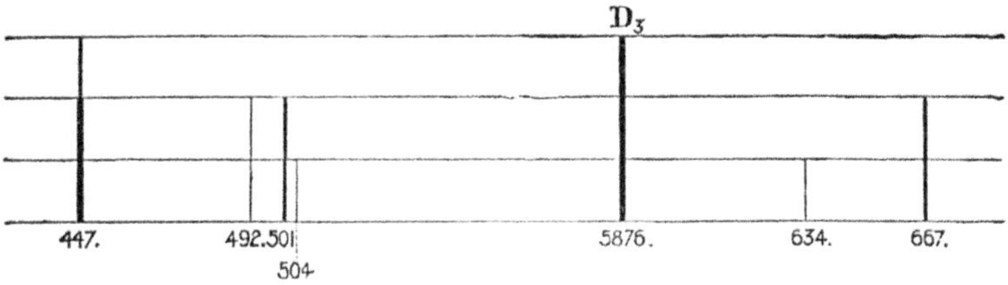

FIG. 22.—Diagram showing order in which lines appear in spectrum of vacuum tube when bröggerite is heated.

It was recorded further that the yellow line was at times dimmed, while the other lines were brightened.

In my second note, communicated to the Royal Society on the 8th May, 1895, I stated that I had never once seen the lines recorded by Thalén in the blue, at λ 4922 and 4715.

It now seems possible that their absence from my previous tubes was due to the fact that the heating of the minerals was not sufficiently prolonged to bring out the gases producing these lines.

It is, perhaps, to the similar high complexity of the gas obtained from cleveite that the curious behaviour of a tube which Professor Ramsay was so good as to send me, must be ascribed. When I received it from him the glorious yellow effulgence of the capillary while the current was passing was a sight to see. But after this had gone on for some time, while the coincidence of the yellow line with D^3 of the chromosphere was being inquired into, the luminosity of the tube was considerably reduced, and the colours in the capillary and near the poles were changed. From the capillary there was but a feeble glimmer, not of an orange tint, while the orange tint was now observed near the poles, the poles themselves being obscured by a coating on the glass of brilliant metallic lustre.

After attempting in vain for some time to determine the cause of the inversion of D^3 and 447 in various photographs I had obtained of the spectra of the products of distillation of many minerals, it struck me that these results might be associated with the phenomena exhibited by the tube, and that one explanation would be rendered more probable if it could be shown that the change in the illumination of the tube was due to the formation of platinum compounds, platinum poles being used. On 21st May I accordingly passed the current and heated one of the poles, rapidly changing its direction to assure the action of the negative pole, when the capillary shortly gave a very strong spectrum of hydrogen, both lines and structure. A gentle heat was continued for some time, and apparently the pressure in the tube varied very considerably, for as it cooled the hydrogen disappeared, and the D^3 line shone out with its pristine brilliancy. The experiment was repeated on 24th May, and similar phenomena were observed.

Some little time after[1] Professors Runge and Paschen, from an entirely different standpoint, arrived at exactly the same conclusion.

The employment of exposures extending over seven hours has given a considerable extension in the number of lines, and the bolometer has been called in to investigate lines in the infra-red; better still, they have employed well-practised hands in searching for series of lines. Operating by chemical means upon a crystal of cleveite free from any other mineral, they have obtained a product so pure that from these series there

---

[1] *Nature*, 26th September, 1895.

are no outstanding lines. Very great weight, therefore, must be attached to their conclusions.

As a result of their investigations, Drs. Runge and Paschen stated that the gas given off even by a pure crystal of cleveite is not simple. In their view the mixture consists of two constituents.

This conclusion was arrived at from the following considerations:—

"The wave-lengths (λ) of the lines belonging to the same series are always approximately connected by a formula somewhat similar to Balmer's

$$\text{“} 1/\lambda = A - B/m^2 - C/m^4.$$

"A determines the end of the series towards which the lines approach for high values of m, but does not influence the difference of wave-numbers of any two lines. B has nearly the same value for all the series observed, and C may be said to determine the spread of the series, corresponding intervals between the wave-numbers being larger for larger values of C. As B is approximately known, two wave-lengths of a series suffice to determine the constants A and C, and thus to calculate approximately the wave-lengths of the other lines. It was by this means that we succeeded in disentangling the spectrum of the gas in cleveite, and showing its regularity.

"In the spectrum of many elements two series have been observed for which A has the same value, so that they both approach to the same limit. In all these cases the series for which C has the smaller value, that is to say, which has the smaller spread, is the stronger of the two. In the spectrum of the gas in cleveite we have two instances of the same occurrence. One of the two pairs of series, the one to which the strong yellow double line belongs, consists throughout of double lines whose wave-numbers seem to have the same difference, while the lines of the other pair of series appear to be all single. Lithium is an instance of a pair of series of single lines approaching to the same limit. But there are also many instances of two series of double lines of equal difference of wave-numbers ending at the same place as sodium, potassium, aluminium, etc. There are also cases where the members of each series consist of triplets of the same difference of wave-numbers, as in the spectrum of magnesium, calcium, strontium, zinc, cadmium, mercury. But there is no instance of an element whose spectrum contains two pairs of series ending at the same place. This suggested to us the idea that the two pairs of series belonged to different elements. One of the two pairs being by far the stronger, we assume that the stronger one of the two remaining series belongs to the same element as the stronger pair. We thus get two spectra consisting of three series each, two series ending at the same place and the third leaping over the first two in large bounds and ending in the more refrangible part of the spectrum. This third series we suppose to be analogous to the so-called principal series in the spectra of the alkalis, which show the same features. It is not impossible, one may even say not unlikely, that there are principal series in the spectra of the other elements. But so far they have not been shown to exist.

"Each of our two spectra now shows a close analogy to the spectra of the alkalis.

"We therefore believe the gas in cleveite to consist of two, and not more than two, constituents."

To the one containing the line D^3, which I discovered in 1868, the name helium remains; the other for the present we may call "gas X."[1]

The chief lines of these two constituents are as follows, according to Runge and Paschen, the wave-lengths being abridged to five figures.

HELIUM.

Principal Series.	1st Subordinate Series.	2nd Subordinate Series.
2663·3	3456·9 ?	3481·6
2677·2	3461·4 ?	3490·8
2696·2	3466·0	3502·5
2723·3	3471·9	3517·5
2763·9	3479·1	3537·0
2829·2	3487·9	3563·1
2945·2	3498·8	3599·5
3187·8	3512·6	3652·1
3888·8	3530·6	3733·0
	3554·6	3867·6
	3587·4	4121·0
	3634·4	4713·3
	3705·2	7065·5
	3819·8	
	4026·3	
	4471·6	
	5875·9	

[1] In the many comparisons I had to make, I soon found the inconvenience of not having a name for the gas which gave 667, 501 and other lines. When,

GAS X.

Principal Series.	1st Subordinate Series.	2nd Subordinate Series.
3176·6	3756·2	3770·7
3196·8	3768·9	3787·6
3211·6	3785·0	3838·2
3231·3	3805·9	3878·3
3258·3	3833·7	3936·1
3296·9	3872·0	4024·1
3354·7	3926·7	4169·1
3447·7	4009·4	4437·7
3613·8	4143·9	5047·8
3964·9	4388·1	7281·8
5015·7	4922·1	
	6678·4	

More recently Professor Ramsay has abandoned his view of the simple nature of the cleveite gas, and states that from his experiments "there appears ground for the supposition that helium is a mixture."[1]

THE EXISTENCE OF THE NEW GASES IN CELESTIAL BODIES.

And now comes the great revelation, and it is this. The majority of the lines classed as unknown in the spectra of the Orion nebula, stars of Group III, and the sun are really due to the cleveite gases.

The following table sets this result out. It will be seen that of seventeen unknown lines, twelve have been run to earth.

therefore, Professors Runge and Paschen, who had endorsed my results, and had extended them, called upon me, I thought it right to suggest to them that, sinking the priority of my own results, we should all three combine in suggesting a name. Professor Runge (under date 20th October) wrote me "the inference that there are two gases in a spectroscopical one, being based on the investigation of the 'series.' Now, though we think this basis quite sound, we must own that the conclusion rests on induction. . . . For this reason we do not want to give a name to 'gas X'." I have so far suggested no name, though Orionium and Asterium have been in my mind.

[1] *Nature*, vol. liii, p. 598.

Comparison of Lines in Orion Nebula and Bellatrix.

Orion Nebula.		Bellatrix and Eclipse 1893.	Origin.
Campbell.	Lockyer.		
3869	*3869 (7)	†3867·5 (Falls	He.
3889	3888 (7)	3888 on He.)	He.
—	4011 (3)	4009 (8)	X
4026	4026 (5)	4026 (10)	He.
—	4042 (1)	4041 (3)	Still unknown
4067	4068 (3)	4070 (3)	Still unknown
4121	4121 (1)	4121·3 (7)	He.
4143	4143 (1)	4144 (8)	X
—	4168 (1)	4169 (5)	X
4265	4270 (3)	4268 (7)	Still unknown
4389	4390 (3)	4389 (8)	X
4472	4472 (7)	4472 (10)	He.
—	4540 (3)	4541 (1)	Still unknown
—	4628 (3)	4630 (3)	Still unknown
4714	4716 (3)	4715 (5)	He.
—	*4924 (5)	†4922·1 (8)	X
5874	5875·8	5875·8	D^3 He.

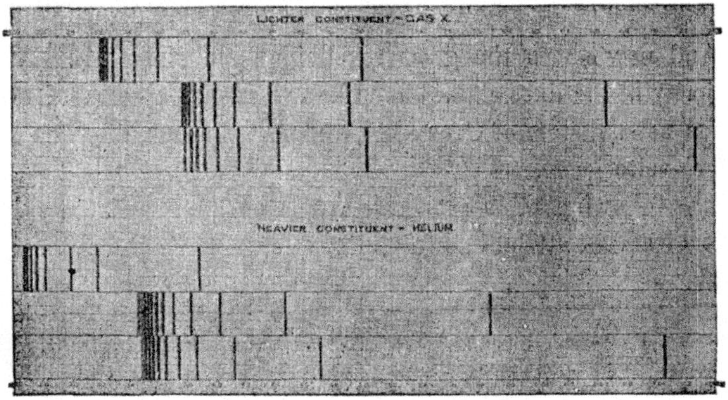

FIG. 23.—Runge and Paschen's results suggesting that cleveite gives off two gases, each with three series of lines.

* Between these λλ there are forty-two lines in the Orion photo of which six are known other than He. and X.

† Between these λλ there are forty-five lines in the Bellatrix photo of which five are known other than He. and X.

The following tables give the complete list of lines and the celestial body in which they have been traced :—

In the tables, under "Sun," C, followed by a number, indicates the frequency as given by Young; E indicates the lines

HELIUM.

	Sun.	Star or Nebula.
11170		
3889	C E	N. III. γ
3188		
2945		
2829		
2764 } *		
2723		
2696		
2677		
2663		
5876	C 100 E	N. Bellatrix.
4472	C 100 E	N. III. γ
4026	C 25 E	N. III. γ
3820	E	III. γ
3705		
3634		
3587		
3555		
3531		
3513		
3499 } *		
3488		
3479		
3472		
3466		
3461		
3457		
7066	C 100	
4713	C 2 E	N. III. γ
4121	E	N. III. γ
3868	?	
3733	E	Bellatrix
3652		
3599		
3563		
3537 } *		
3517		
3503		
3491		
3482		

photographed during the eclipse of 1893. Under "star or nebula," the references are to the tables given in my memoir on the nebula of Orion (*Phil. Trans.*, vol. clxxxvi, 1895, p. 86 *et seq.*). N = Nebula of Orion.

GAS X.

	Sun.	Star or Nebula.
5016	C 30 E	
3965	?	III. γ
3614 ⎫	E	
3448 ⎪		
3355 ⎪		
3297 ⎪		
3258 ⎬*		
3231 ⎪		
3212 ⎪		
3197 ⎪		
3177 ⎭		
6678	C 25	
4922	C 30 E	N. III. γ
4388	E	N. III. γ
4144	E	N. III. γ
4009		III. γ
3927		Bellatrix
3872		Bellatrix
3834	E	Hid by H. line
3806		Bellatrix
3785*		
3769*		
3756*		
7282		
5048	C 2	
4438		Bellatrix
4169		Bellatrix
4024	?	
3936	Hid in K.	
3878	C E	α Cygni
3838	C E	α Cygni
3788*		
3771*		

* Means that these lines are out of the range of my observations.

The annexed reproduction of a photograph of Bellatrix will show how striking has been the result of the discovery so far as stellar spectra are concerned.

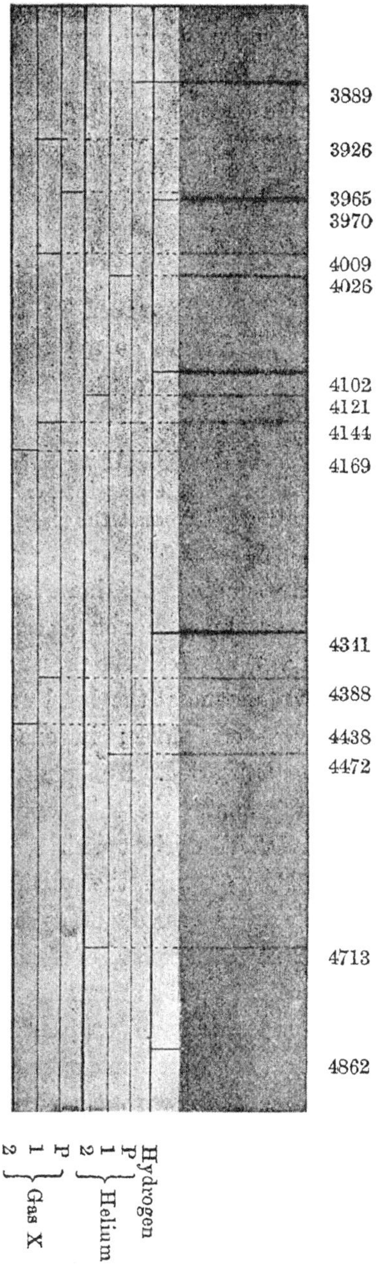

FIG. 24.—The spectrum of Bellatrix showing the lines of hydrogen and those which have been traced to the three series of lines in the spectrum of the gases obtained from minerals.

F

Hydrogen, helium, and gas X are thus proved to be those elements, which are, we may say, completely represented in the hottest stars and in the hottest part of the sun that we can get at. Here then, in 1897, we have abundant confirmation of the views I put forward in 1868 as to the close connection between helium and hydrogen.

Effects of Diffusion.

A diffusion experiment described in their paper enabled Messrs. Runge and Paschen to go a stage farther, and to announce that of their two constituents the gas giving D^3 was the heavier one. They also added :—

> "From the fact that the second set of series is on the whole situated more to the refrangible part of the spectrum, one may, independently of the diffusion experiment, conclude that the element corresponding to the second set is the heavier of the two."

As they themselves pointed out, however, the result was not final, because the pressures were not the same. I have recently made some experiments in which the pressures remain the same.

A ∩-tube was taken, and at the bend was fixed a plaster of Paris plug about 1·5 cm. thick; in one of the limbs two platinum wires were inserted. The plug was saturated with hydrogen to free it from air; the tube was then plunged into a mercury trough, and fixed upright with the limbs full of mercury. Into the leg (A) with the platinum wires a small quantity of hydrogen was passed, and as soon after as possible another small quantity of a mixture of helium and hydrogen from samarskite was put up the other limb (B) of the ∩-tube.

Immediately after the helium was passed into the limb (B) spectroscopic observations were made of the gas in the limb (A); D^3 was already visible, and there was no trace of 5015·7. This result seems to clearly indicate that if a true diffusion of one constituent takes place, the component which gives D^3 is lighter than the one which gives the line at wave-length 5015·7.

Although this result is opposed to the statement made by Runge and Paschen, it is entirely in harmony with the solar and stellar results.

In support of this I may instance that of the cleveite lines associated with hydrogen in the chromosphere and stars of Group IIIγ, those allied to D^3 are much stronger than those belonging to the series of which 5015·7 forms part.

Minerals Examined.

So far I have worked upon some seventy minerals, and I have found the yellow line in sixteen. The following are the minerals, etc., which have been investigated; those which give the D^3 line being marked with an asterisk:—

*Æschynite.
Almandine.
Anglesite.
Anhydrite.
Augite.
Barytes.
*Bröggerite.
Bronzite.
Calco-uranite.
Cassiterite.
Celestine.
Chalk.
Charnockite.
Chromite.
*Cleveite.
Columbite.
Crocidolite.
Cupro-uranite.
*Eliasite.
Enstatite.
*Euxenite.
*Fergusonite.
Franklinite.
Gadolinite.
Gahnite.
Geikielite.
Gneiss.
Granite.
Graphite.
*Gummite.
Hæmatite.
*Hielmite.
Hornblende.
Hypersthene.
Ilmenite.
Iridosmine.
Kielhanite.
Kyanite.
Ludwigite.
Magnesium.
Magnetite.
Manganese Nodule.
Minium.
*Monazite.
Obsidian.
Olivine.
Olivine-Enstatite.
*Orangeite.
Orthite.
Pitchblende.
Plumbic Ochre.
*Polycrase.

*Pyrochlore.
Quartz.
Red Clay.
Rhodonite.
*Samarskite.
Schorlomite.
Sphene.
Staurolite.
Thorite.
*Thoro-gummite.
*Uraninite.
Uranocircite.
Uranophase.
Wulfenite.
Wolfram.
Xenotine.
*Yttro-Gummite.

CHAPTER VI.—THE RELATION OF STARS TO NEBULÆ.

THE OLD VIEWS.

THE various views which have been held time out of mind with regard to the relation of stars to nebulæ, and the special nature of both these classes of celestial bodies have always had the greatest interest for mankind, for those at all events among us who like to know something about the universe in which our lot is cast. No dividends, unfortunately or fortunately, depend upon the discussion or even the application of any branches of inquiry which are necessary in order to make progress along the lines of thought thus opened up; scant attention is paid to them by educational bodies, for they lead to no profession. Still, in spite of this, some of the noblest triumphs of the human mind have been made in that region where man finds himself face to face with the mysteries of the distant heavens, and indeed no chapter in human history is more interesting than that in which we read how man has struggled with the mysteries which surround him in the depths of space.

To consider completely the Sun's place in Nature the relation of these two apparently different classes of celestial bodies, to which I have referred, must be gone into. Thanks to the advance of modern science, pictures of clusters of stars and of nebulæ, in which we see those bodies very much better than we can in the best equipped observatories in the world, are at our disposal, for it so happens that the enormous progress which has recently been made in the application of photography to astronomical work enables us to get permanent records of parts of them which are so dim that they never have been and never will be directly revealed to the eye of mortals.

It will be well to call attention to some typical examples both of clusters and nebulæ. This I am enabled to do by the kindness of Dr. I. Roberts, to whose instrumental means I have already called attention. A condensed star cluster is well represented by the one in Libra, which well indicates that in the case of a star cluster it is obviously a case of separate stars gradually congregating towards the centre.

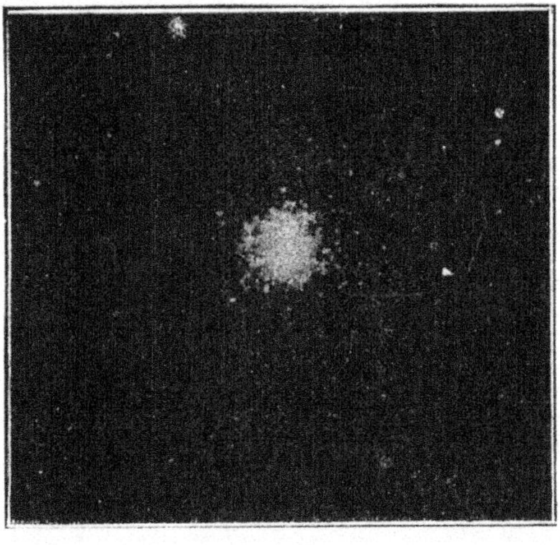

FIG. 25.—The Cluster 15 M Libræ, from a photograph by Dr. Roberts.

In the photograph of the great nebula in Andromeda, which is given as a term of comparison, there is a filmy sort of luminosity, which is quite distinct from the neighbouring stars. To bring out the transforming effect of prolonged exposures on the nebulæ I give two views of the nebula of Orion which show

FIG. 26.—The Great Nebula in Andromeda, from a photograph by Dr. Roberts.

that the nebula which we see ordinarily in our telescopes is only a very, very small fraction of the real nebula as it really exists, when we can get at it under the best possible observing conditions. On the left of the figure we have the results when it has been photographed by means of a telescope powerful

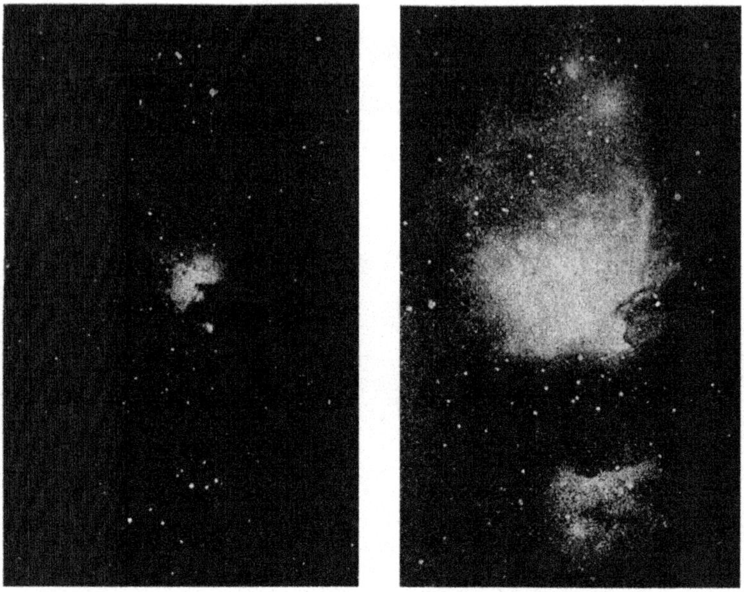

FIG. 27.—Orion nebula photographed with short and long exposures.

enough to give us the brighter portions. On the right we have another photograph of the nebula exactly on the same scale, in which the nebula that we usually see occupies only a very small place; the only difference between the two photographs is that one has been exposed for a very long time to enable us to fix and to study the very dim reproduction of certain parts of it, whilst the first one was exposed only for a short time, in order that we might dwell effectively on that part only of the nebula which is generally visible to the human eye with an ordinary telescope.

A rapid glance at these photographs will indicate that there is a very great difference in the appearance of star-clusters and nebulæ, and when we compare these two great groups of celestial bodies with the greatest care we find that, at all events in appearance, there is an enormous difference between them; that a nebula is certainly unlike an ordinary star-cluster. This is so obvious that even those who first observed the very few nebulæ which are visible to the naked eye, such as that in Orion, were thrown into the greatest wonderment by their strange appearance.

Let us go back 150 years. Here is what the French philosopher Maupertuis said about them in the year 1745.

"The first phenomenon is that of those brilliant patches in the sky which are named nebulæ, and which have been considered as masses or groups of small stars; but our astronomers, with the aid of better telescopes, have only seen them as great oval areas, luminous and with a light brighter than the rest of the heavens. Huygens first discovered one in the constellation of Orion; Halley, in the *Philosophical Transactions*, pointed out six, the first in the sword of Orion, the second in the constellation of Sagittarius, the third in the Centaur, the fourth before the right foot of Antinöus, the fifth in Hercules, and the sixth in Andromeda. Five of these spots having been observed with a reflector of 8 feet, only one of them, the fourth, could be taken for a group of stars; the others seem to be great shining areas, and do not differ among themselves, except that some are more round and others more oval in shape. It seems also that in the first the little stars which one discovers with the telescope are not capable of causing this brightness. Halley was much struck with these phenomena, which he believes capable of explaining a thing which seemed difficult to understand in the Book of Genesis, viz., that light was created before the sun. Durham regards them as holes through which one discovers an immense region of light, and finally the empyrean heaven itself. He professes to have been able to distinguish that the stars which are seen in some of them are very much less distant from us than the spots of light themselves." [1]

Hence we see that 150 years ago some of our keenest intellects were struggling with the questions, involved in mystery, which had been started by the discovery of these nebulous bodies in space. That was in the year 1745. Soon after this, in the year 1755, Kant, who was a German, though he was a Scotchman by descent, brought out an hypothesis in which he attempted to show that there was the closest possible connection between stars and the clusters and nebulæ of which

[1] *Discours sur les différentes figures des Astres*, chap. vi. pp. 104–14.

Maupertuis wrote. He held distinctly that the stars were produced by some action brought about in nebulæ; in other words, that the nebulæ represented a first stage out of which stars, representing a later stage, were produced by certain processes of evolution.

From 1755 we pass to 1796, at which date we find a great Frenchman, Laplace, practically rediscovering and reasserting the same thing. It is believed that he knew nothing of Kant's prior work, and therefore we have the advantage of dealing with the results of the thoughts of two great minds. Laplace came to the same conclusion as Kant, so far as it went, but he went further than Kant did, because he held that the nebulæ really represent enormous masses of elastic gas at a very high temperature, and that therefore the stars, which he conceived, as Kant had conceived, to be produced by evolutionary processes from these nebulæ, were really produced from incandescent masses of gas.

Now, seeing that our sun is a star, it is perfectly clear from this that both Kant and Laplace agreed that the sun, representing a star, had originally been produced from a nebula.

About the time of Laplace, *i.e.*, about 1796, Sir William Herschel was making England famous by the discoveries rendered possible by that wonderful telescope which he had erected at Slough. There, for the first time, the possible similarities and the possible differences of these two great groups of celestial bodies were subjected to the most minute and laborious scrutiny. He came absolutely to the same conclusion as his predecessors had done, and for Sir William Herschel there was no doubt whatever that from the most irregular nebula to the densest star there was a gradual process of change; that there was no radical difference, but that the star represented simply the result of certain evolutionary changes. This view thus strengthened held the field for some years; then a larger telescope was made by Lord Rosse; a 6-foot mirror was now available instead of the

4-foot one which had been erected by Herschel at Slough. Lord Rosse—we find the whole story admirably told in Professor Nichol's book, *The Architecture of the Heavens*—came to the conclusion that when he observed a so-called nebula on the finest possible nights, when the air was stillest, and the magnifying power which he could use was greater than usual, he could see what he called the possibility of a resolvability in it. That is to say, nebulæ might after all really be star-clusters, only immensely remote, so that the light of all the stars was, as it were, so blended together as to give that appearance of a candle seen through horn, which Maupertuis and his predecessors had observed.

Next we come to the year 1862, and we find a new instrument brought to bear, which at once drove into thin air all the statements which had been made on what had turned out to be a line of inquiry which was incapable of giving a final verdict. It so happened that in that year there was a very powerful combination formed by a distinguished chemist and philosopher, Dr. William Allen Miller, the Treasurer of the Royal Society, who had already done admirable spectroscopic work, and a neighbour of his, Mr. Huggins, who had mounted a powerful telescope in 1856. The spectroscope, which was then practically a new instrument, was attached to the telescope.

The first question put to the combined instruments was: What is starlight like? It was found that the stars give a spectrum very much like the spectrum of the sun, in some cases at all events, and that this spectrum could be defined in the light of the third axiom (p. 15), that part of the light was absorbed, there were dark lines in the spectrum; and thus we knew that light had been absorbed by an atmosphere surrounding something which was very much hotter than itself, and in that way the science of solar and stellar physics was founded.

The second question put to this instrument was what is the light of the nebulæ like?

I have already stated that Laplace held that in these bodies we were dealing with gas at a high temperature. From the time of Tycho Brahe downwards people had an idea that the nebulæ were "fiery." What should we expect to get in our instrument?

The second axiom tells us that, if we are dealing with matter in a state of gas, or anything vapourous at very high temperature, we shall get bright lines only. The question as to the nebulæ was put in 1864, and, curiously enough, when the observation was made, the observer, Dr. Huggins remarked: "At first I suspected some derangement of the instrument had taken place, for no spectrum was seen, but only a short line of light perpendicular to the direction of dispersion."[1] "Only a line" was exactly what I suppose Laplace would have given all he possessed to see, if spectrum analysis had been invented in his day.

That line settled the question. There was certainly a tremendous spectroscopic difference between stars and nebulæ, and this difference has been emphasised by subsequent researches. It is evident, therefore, that Lord Rosse's suspicion that the nebulæ might, after all, be found to be resolvable into star clusters when greater optical power was used, was proved to be erroneous.

This, then, was a great point gained—a solid advance, but unfortunately the spectroscopic observations of the nebulæ led Dr. Huggins to the conclusion that the result of his inquiries was rather to show that the connection, which had been asserted both by Kant and Laplace, and which had been accepted by everybody up to then, really did not exist. In a paper which detailed these spectroscopic observations, published in 1864, Dr. Huggins stated his conclusion that the nebulæ, instead of having anything whatever to do with any evolutionary line along which both nebulæ and stars might be traced, possessed a

[1] *Phil. Trans.*, 1864, p. 438.

structure and a purpose in relation to the universe altogether distinct and of another order.

He wrote :—

"We have in these objects to do no longer with a special modification only of our own type of suns, but find ourselves in the presence of objects possessing a peculiar and distinct plan of structure." [1]

And again in 1865 :—

"The nebulæ which give a gaseous spectrum are systems possessing a structure, and a purpose in relation to the universe, altogether distinct and of another order from the group of cosmical bodies to which our sun and the fixed stars belong." [2]

So that the first apparent teaching which we got from the spectroscope practically put us in a very considerable difficulty; if it were accepted, of course the views of Kant and Laplace would have to be rejected.

Now, if we form any conception of nebulæ changing into stars, we begin by knowing that the stars are very much denser than the nebulæ—taking the sun as an instance, the star practically close to us—and that, as the stars are denser than the nebulæ from which they are formed, they must at some stage be hotter than the nebulæ, instead of being colder.

This conclusion follows upon the application of thermodynamics, and had been indicated in 1854 by Helmholtz, who showed that—

"It was not necessary to suppose the nebulous matter to have been originally fiery, but that the mutual gravitation between its parts may have generated the heat to which the present high temperature of the sun is due."

THE NEW VIEW.

Sir William Thomson, now Lord Kelvin, pointed out quite distinctly, in 1871, that the hypothesis of fiery nebulous matter —by that meaning nebulous matter hotter than the stars—was invented before the discovery of thermodynamics.

---

[1] *Phil. Trans.*, 1864, p. 412.

[2] *Proc. Roy. Soc.*, 1865, vol. xiv, p. 39.

"The old nebular hypothesis supposes the solar system, and other similar systems through the universe, which we see at a distance as stars, to have originated in the condensation of fiery nebulous matter. This hypothesis was invented before the discovery of thermo-dynamics, or the nebulæ would not have been supposed to be fiery; and the idea seems never to have occurred to any of its inventors or early supporters that the matter, the condensation of which they supposed to constitute the sun and stars, could have been other than fiery in the beginning."[1]

More than this, Lord Kelvin told us how he could imagine a condition of nebulæ which might ultimately condense into stars without violating the laws of thermodynamics, which were completely traversed by Laplace's view; and he referred to a suggestion that had been made by Professor Tait, who supposed that the luminosity of nebulæ, and even the spectroscopic appearances which have been observed, might be explained by supposing that we were dealing with gaseous exhalations proceeding from the collisions of meteoric stones; and he also pointed out that possibly that would not only explain the luminosity of nebulæ, but the luminosity of comets as well.

"But a solution, which seems to me in the highest degree probable, has been suggested by Tait. He suppoees that it may be by ignited gaseous exhalations proceeding from the collision of meteoric stones, that nebulæ and the heads of comets show themselves to us."[2]

When I commenced my general survey in 1887, Dr. Huggins' view, in spite of the caveat of Lord Kelvin, was supposed to hold the field, and hence it was generally imagined that the observations of Dr. Huggins justified the idea that the nebulæ were masses of incandescent permanent gases.

It may well be imagined that the view that the nebulæ consisted of one or two permanent gases at once led to several most remarkable views of the general constitution of the heavens.

Towards the end of this nineteenth century chemists claim to know something of the materials which have built up the

[1] *Brit. Assoc. Report*, 1871, p. 99.

[2] Lord Kelvin, *Brit. Assoc. Address*, 1871.

planet on which we dwell, and if we consult any of the books which have been written on spectrum analysis, giving the results of the work during the last thirty years or so, we find it stated over and over again that the spectroscope has put it for ever beyond doubt that the chemistry of the skies, *i.e.*, the chemistry of the various bodies in space, which are at a sufficiently high temperature to enable us to examine them spectroscopically, exactly resembles the chemistry of the earth. So that, if this were true, we should have a common chemistry of the earth, of the stars, and among the stars of course our own sun.

On the other hand, we should have, according to Dr. Huggins, absolutely and completely distinct from these bodies another class, the nebulæ, in which the chemistry is absolutely and completely different. This was so clearly the idea suggested to philosophical students of these questions, that Dr. Wolf, a famous French astronomer, who has written an all-important book for those who are interested in these inquiries, *Les hypothèses cosmogoniques*, published in 1886, writes:—

"If we admit the data of spectrum analysis as to the gaseous state of these singular bodies, the nebulæ, and the simplicity of their composition, one is led to see in them only the residuum of the primitive matter after condensation into suns and into planets has extracted the greater part of the simple elements which we find on the earth and chemically in some of the stars."[1]

It was perfectly clear then to Dr. Wolf that, if the constitution of the nebulæ was anything like what was supposed to have been revealed by the early spectroscopic observations, we were dealing with a residuum. There was one kind of action in space, bringing together one class of elements with which we are familiar here, and forming them into stars, suns, and planets; but there was another kind of matter which declined to form part of these aggregations, which remained by itself, and finally put on the appearance of the so-called nebulæ.

[1] *Hypothèses Cosmogoniques*, 1886, p. 7.

The complete discussion of all observations of comets, nebulæ, and stars available in 1877 led me to the conclusion which Professor Tait had already suggested in the absence of spectroscopic evidence; and the view that the nebulæ may be explained apart from any fiery permanent gas, and that we have simply to look to a meteoritic origin to explain both the appearances and the spectrum was shown to be in accordance with all the known facts.

Dealing with nebulæ, then, as a whole, it does not seem too much to say that we are justified in supposing that they may advance towards condensation along two perfectly distinct lines. If we consider a regular spiral nebula, like the one in Andromeda, or a planetary nebula, we may imagine them living their life as nebulæ without very much disturbance; there is not much fighting to be done, they progress in orderly fashion towards the condition of complete condensation at the centre.

But there is another way.

In the nebula of Orion we get absolute absence of anything like regularity. In any part where the structure can be studied, we find it consists of whirls and streams crossing each other, some of them straight, some of them curved, the whole thing an irregular complicated mixture of divergent movements, so far as the photographs, which are absolutely untouched, can give us any idea of what is going on. Take, for instance, the magnificent streamer trending upwards. It gradually becomes brighter until it reaches one of the brightest parts of the nebula; and observe, also, the stars which seem dotted over it as on a shield. It is quite obvious that we cannot, in such a structure as that, expect to get the same conditions that we met with in the nebula of Andromeda, and in the planetary nebulæ. And, in fact, we do not. In this nebula, which speaks of disturbance in every inch of it, we have considerable differences in the spectroscopic indications. Carbon is replaced by hydrogen and helium. In such a nebula as this, it is impossible for us to pick out the place of condensation; the condensation may be held to be anywhere, for disturbances are obviously everywhere. And we must remember that the part of the nebula ordinarily seen is but the brightest part of a nebula extending over a space in the surrounding neighbourhood, which recent research shows is scarcely limited to the constellation itself.

The arguments I chiefly employed to prove the connection between stars and nebulæ was the great similarity of the spectra of nebulæ and of the bright-line stars; the relationships between the spectra of comets and nebulæ; the evidences of a relatively low temperature in the latter, and the wave-length and appearance of the chief nebula line.

Shortly after I brought this evidence together photography came to our aid, and I am so fortunate as now to be able to prove the truth of my position by many appeals to Nature herself; that is, I can now refer for demonstration, not only to spectroscopic evidence, but to autobiographical records with which the heavens themselves have supplied us. I need only refer to one instance in this place. Among the finest and most wonderful of the nebulæ is one which, unfortunately, we do not see here, because it is in the southern hemisphere; it is that surrounding the star η in a brilliant constellation, Argo, which it is quite worth while to go south to see, were there no other reasons. From the photograph we see that there is such an intimate connection, such an obvious relation, between star and nebula, that it is impossible for us to imagine for one moment that they are not most closely and intimately connected.

The discussion of all the observations available from these and other points of view which need not be detailed here showed, beyond all question, that there was no real ground for supposing a great difference between the nebulæ and the stars. In the year 1887, after testing the views on this question by an appeal to all the available observations, I stated that the facts taken in all their generality showed that the nebulæ simply

FIG. 28.—Nebula round η Argus (Dr. Gill).

represent early stages of evolution; that is to say, that we have a continuous and orderly progression from the nebulæ to the oldest star, and that the nebulæ represent the first stage, and the oldest star or planet represents the last. It seemed to be perfectly clear from the discussion that we were justified in stating that every nebula which is visible now will some time or other, owing to the condensation of its various parts, become a star of some order or another; and that it is equally true to say that every star which we see now in the heavens, knowing it to be a star, has really been a nebula at some time or another.

It was the discussion of the cometary phenomena which indicated that the beginnings of nebulæ consist in meteoritic swarms, and indeed the complete inquiry showed that these meteoritic particles might account equally well, as Professor Tait had suggested, both for the luminosity of comets and of nebulæ.

How the New View has Fared.

How, then, has the new hypothesis fared with regard to this point? I am glad to say that among the first to accept the new evidence, proving that nebulæ are really early stages of evolution of stars, was Dr. Huggins himself, the observer whose directly opposed statement I have quoted. He says now not only that these bodies may represent early forms, but he places them in the line of evolution where I had placed them. His exact words are as follows:—

> "It may be that they represent an early stage in the evolutionary changes of the heavenly bodies." . . . .
>
> "These bodies may stand at or near the beginning of the evolutionary cycle, so far as we can know it. They consist probably of a gas at a high temperature and very tenuous."[1]

He even adduces the same evidence which I had brought forward in several of the arguments which I had employed. Dr. Huggins made a reference to this question as President of the British Association in the year 1891. That part of his address is really an argument in favour of the views that I have been insisting upon since 1886, and his agreement seems all the more important, since Dr. Huggins appears to have arrived at these conclusions quite independently. Not one word is said throughout the address of any arguments which I had used, or of any line of thought or observation on which I had founded the various statements which I had made; and, therefore, it would be charitable to suppose that he was unacquainted with my work when that address was given to the world. I think I am quite justified in drawing general attention to this

[1] *Proc. Roy. Soc.*, vol. xlvi, 1889, p. 59

very extraordinary change of opinion; so extraordinary, indeed, was it that it is clear that Dr. Huggins felt that it was of importance to himself that the change should be explained; and he confesses in the address to which I refer that the communication he made to the Royal Society in 1864 was not entirely founded on scientific evidence, but partly made under, to use his own words, "the undue influence of theological opinions then widely prevalent."

It is fortunate for that large public which takes an interest in scientific matters, that the statements made by men of science—on which statements it is compelled to rely, and on which it builds its views of the universe and all it contains—are not, as a rule, thus unduly influenced by prevalent theological opinions.

I do not think I am going too far in stating, that this fundamental point of the meteoritic hypothesis is now conceded on all hands; indeed, it has arrived at that interesting stage in which the only objection urged against it is that it is not new.

The more one knows of the history of human thought, especially during the last three centuries, the more important does this new acknowledgment of the working of evolution seem. Indeed, it is one of the most important truths established during the present century.

CHAPTER VII.—THE CHEMISTRY OF THE NEBULÆ.

The First Spectroscopic Results.

So far, then, we have found that there is now a common agreement that nebulæ represent early forms in the evolution of the heavenly bodies.

It becomes important to point out that some of the first statements made on spectroscopic evidence regarding their chemistry must now be withdrawn as completely as those relating to their relation to the other bodies composing the cosmos.

Dr. Huggins' view as to their chemical constitution was that they were mainly, or to a large extent, composed of nitrogen. This idea was based upon his measurement of the chief line in the nebular spectrum. When nitrogen is observed by means of a spectroscope, a double line is seen very nearly coincident with the line of the nebulæ. Dr. Huggins thought that the chief nebular line was exactly coincident with one constituent line of the double line of nitrogen, and, because there was apparently no line in the nebular spectrum corresponding with the other, he thought, also, that the nitrogen might not be nitrogen like that with which we are familiar, but an unknown form of it.

He wrote:—

"The speculation presents itself, whether the occurrence of this one line only in the nebulæ may not indicate a form of matter more elementary than nitrogen, and which our analysis has not yet enabled us to detect."[1]

There was no doubt from the beginning that the other line was a line of hydrogen, although there was some slight doubt as to whether the hydrogen in the nebulæ behaved exactly like hydrogen on the earth.

When I undertook the general discussion of all existing spectroscopic observations of the nebulæ in 1887, I brought together the work of others, in the first instance; I made no observations myself. After my book was written, however, I took steps to make special observations of my own, and I propose to refer to these before I go into further details concerning the discussion.

Special Study of the Nebula of Orion.

The nebula of Orion was selected, and photographs were obtained of its spectrum in order that it might be very carefully studied from the point of view of the chemical substances which may be building up this special spectroscopic type.

The photographs were taken with my 30-inch reflector at Westgate-on-Sea in 1890. The focal length of this telescope is about 11 feet, so that, neglecting the loss of light due to absorption in the case of the refractor, and to reflection in the case of the reflector, the brightness of the image formed on the slit of the spectroscope by the Westgate telescope is about sixteen times that of the image formed by the Lick telescope, and it is scarcely necessary to add that, having this great illuminating power, the collimator of the spectroscope has been designed to take full advantage of it.[2]

---

[1] *Proc. Roy. Soc.*, Sept., 1864, p. 444.

[2] The remarks of Professor Keeler (*Ast. and Ast. Phys.*, January, 1894, p. 61) and Mr. Campbell (*Ast. and Ast. Phys.*, May, 1894, p. 385), as to the relative efficiency of telescopes in regard to the observation of spectrum lines, seem to indicate that the matter has not been sufficiently thought out; but it

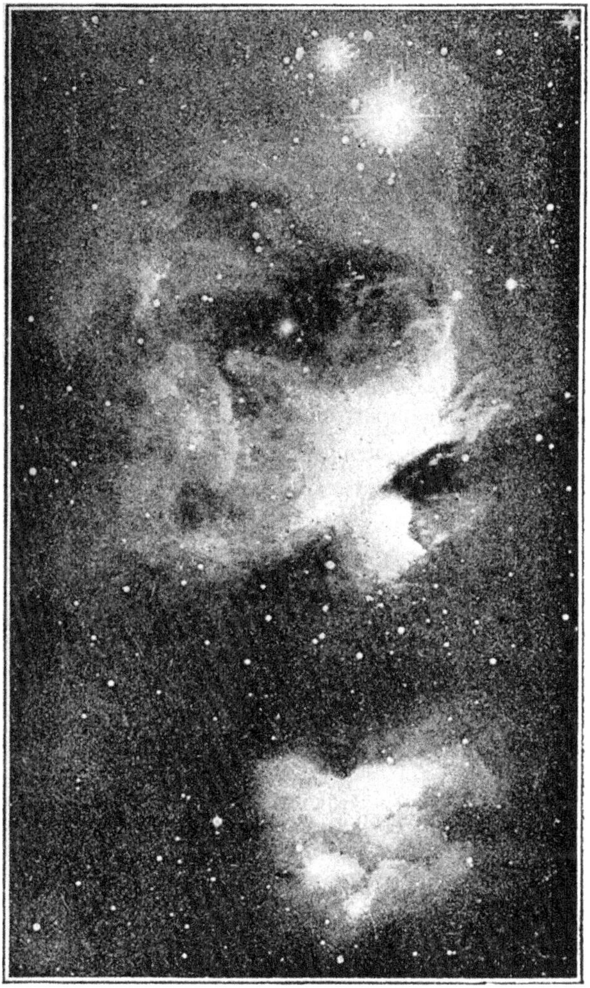

FIG. 29.—The Great Nebula in Orion, from a long exposure photograph by Dr. Roberts.

would appear that Professor Campbell, who has succeeded Professor Keeler at the Lick Observatory, is of the same opinion as myself, for he writes (*Ast. and Ast. Phys.*, 1893, p. 53): "The 36-inch telescope presents several positive disadvantages. . . . . The ratio of the focal length 19 : 1 is much larger than exists in small telescopes, and hence the latter would form much brighter images on the slit plate."

A spectroscope by Hilger, having one prism of 60° and two half-prisms of 30° was used; but, in spite of this feeble dispersion, exposures of four hours were required.

The photographs show a greater number of lines than others which have been described; one contains something like fifty lines, which have been measured; but in the attempt to enlarge, a great many of these have been left behind.

The photographs were taken on February 2, 8, 9, 10, and 11, the fourth with an exposure of three hours.

As a collimator had not been fitted to the tube of the reflector, the exposure of the plate to the flame of burning magnesium, in order to obtain a comparison spectrum, was made by closing the mirror cover, and burning magnesium at its exact centre. One half of the slit was exposed to the nebula, and the other half to the burning magnesium.

The part of the nebula photographed was the region of the trapezium, the brightest part in the accompanying reproduction. In some photographs, in consequence of clock irregularities, the stars of the trapezium have imprinted their spectra upon the plates, but these in no way interfere with the spectrum of the nebula, since a longish slit was used, and the spectra of the stars are narrow.

There is a remarkable and almost absolute similarity between

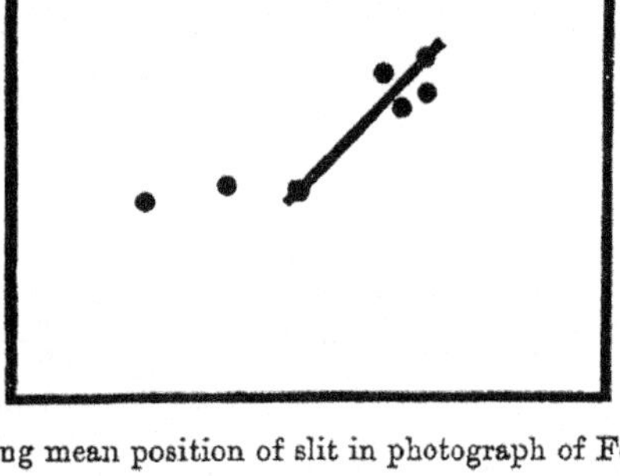

FIG. 30.—Showing mean position of slit in photograph of February 10, 1890.

the photographs obtained. The best one, taken on February 10, shows all the lines of the other photographs with others in addition, and this has, therefore, been selected for the determination of wave-lengths.

The probable mean position of the slit during the three hours' exposure of this photograph is shown in the figure, but the irregularities in the driving caused all the stars in the trapezium to cross the slit at different times.

It has not been found possible to reproduce the negative with advantage in consequence of its small size, but Fig. 31 gives a good idea of the appearance of the eleven principal lines, and the position of the stellar spectra on the plate.

The principal lines are the three ordinarily seen in the visible spectrum, the lines of hydrogen at H_γ, H_δ, H_ϵ, and H_ζ, and the strong line in the ultra-violet near λ 373. H_γ is by far the strongest line in the spectrum. The photograph was measured with a micrometer reading to 0·00001 inch.

Variations of the Spectrum in Different Regions.

In observations with a 12-inch mirror in the year, 1889,[1] "I obtained momentary glimpses of many bright lines between H and H_γ, on October 31." These were also seen by Mr. Fowler, and it was observed that, as the slit was swept across the nebula, in some parts the lines were seen together, while in other parts first one group and then another made their appearance. In the same paper I referred also to the variations *in the same field of view* of some of the lines. These observations were made with an enlarged form of pocket spectroscope, with a dispersion that does not split D. I found that in certain parts of the nebula, in the same field, certain lines were knotted, as often seen in prominences on and off the sun, and in other parts broken; in the former case, whilst the F line thickened equally

[1] *Proc. Roy. Soc.*, vol. xlviii, p. 195.

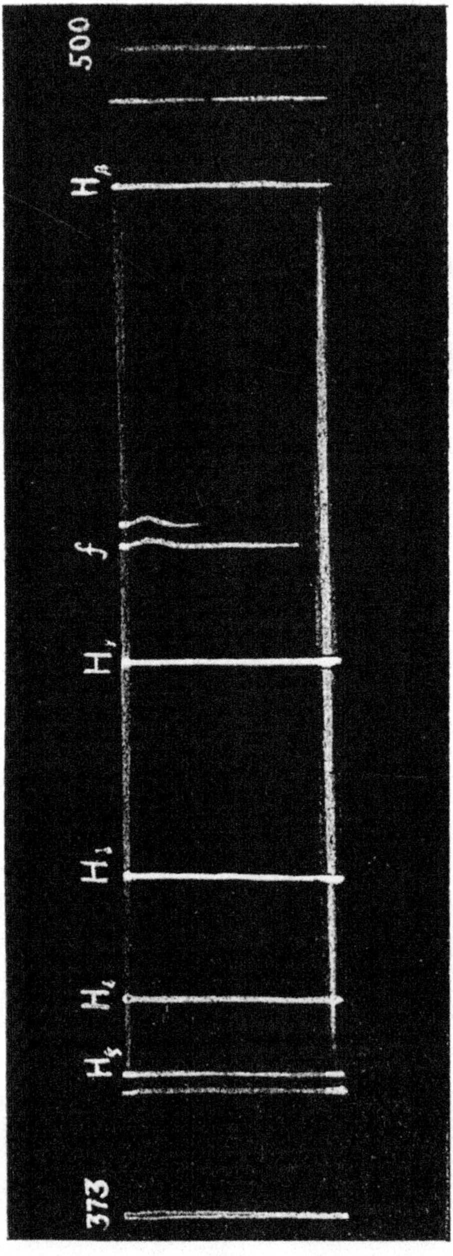

FIG. 31.—Diagram showing the principal lines in the photograph of the spectrum of the Orion nebula, February 10, 1890, with their relative intensities. The spectra of the stars in the trapezium are shown at the bottom of the diagram, while that at the top is the spectrum of the star Bond 685.

on both sides, the chief nebular line thickened only on the more refrangible side.

This result is shown in Fig. 32.

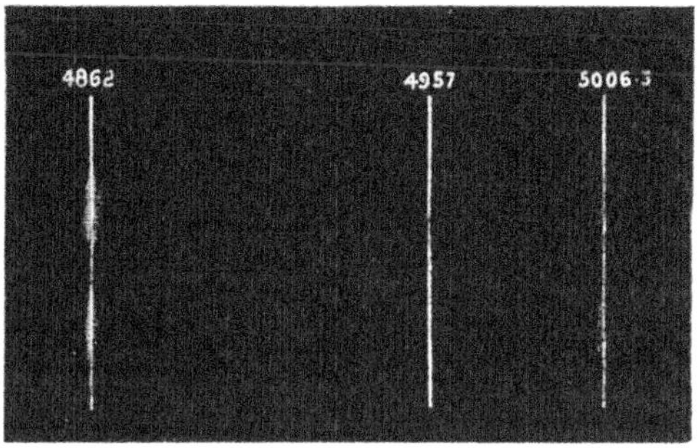

FIG. 32.—Difference in the appearance of the lines at 4862 (F) and 5006·5.

This was confirmed by Messrs. Fowler and Baxandall at Kensington, with the 10-inch equatorial, on October 31 and November 1.

In my own observations in 1891, with the 30-inch reflector at Westgate, the variations were very striking.

One previously recorded is that of the strong line in the ultra-violet near λ 373. This was the strongest line in the photograph taken by Dr. Huggins in March, 1882,[1] but it was not shown in Dr. Draper's photograph taken in the same year.[2]

Dr. Draper wrote:—

"I have not found the line at 3730, of which he (Dr. Huggins) speaks, though I have other lines which he does not appear to have photographed. This may be due to the fact that he had placed his slit on a different region of the nebula, or to his employment of a reflector and Iceland spar prism, or to the use of a different sensitive preparation. Nevertheless, my reference spectrum extends beyond the region in question."

A later photograph (1889), taken by Dr. Huggins,[3] did not show the line in question, the slit being placed on a different

[1] *Proc. Roy. Soc.*, vol. xxxiii, p. 427.

[2] *Amer. Journ. of Science*, (3), vol. xxiii, p. 339.

[3] *Proc. Roy. Soc.*, vol. xlvi, p. 41.

part of the nebula. As already stated, the line is one of the strongest in my photographs, though it is not quite as strong as H_γ. The spectrum photographed by Dr. Huggins, in 1889, differed entirely from those photographed by him in 1882 and 1888, the slit being again placed on a different region of the nebula.

My own photographs are specially interesting, as they indicate differences even in the small area of the nebula which is covered by the slit during a single exposure. Some of the more important variations are indicated in Fig. 31. The stars, the spectra of which are registered on the plate, will be readily identified by a comparison of Figs. 30 and 31, the spectra of the trapezium stars being shown at the bottom of the diagram, and that of the star G. P. Bond 685 (Herschel's ϵ) at the top.

It will be seen, for example, that the line near λ 495 falls off in intensity about the middle of its length, while the lines of hydrogen show no such reduction in the same part of the nebula. If we first consider the phenomena, in the neighbourhood of the star G. P. Bond 685 (Herschel's ϵ), near the trapezium, it will be seen that here the lines 4471 (f) and 4495 are most intense. In this region there is also a distortion of the two lines at 4471 and 4495: they are sharply bent towards the red end of the spectrum, whilst the other lines remain straight. Unfortunately, the spectrum of this star is only shown on the photograph of February 10, and, in the absence of other photographs, it is possible that the displacement of the two lines in question may be due to a distortion of the gelatine film. The displacement of the lines, if real, would indicate a velocity of about 200 miles per second, in the line of sight. Both lines are brightest where they are most disturbed.

It will be seen, also, that where the lines of the nebula cross the continuous spectrum of the star, they are considerably broadened. This is seen in all the principal lines from λ 373 to λ 495.

Where the chief line (500) crosses the spectrum of the star, there is a decided indication of a reversal. As it approaches the star, the line bifurcates and reunites on the other side, leaving a short dark line where it crosses the spectrum of the star, as shown in Fig. 31. This reversal is not seen in the case of the hydrogen lines, but if it be subsequently confirmed in the case of 500, it will be an indication that some of the nebulous matter lies in front of the star in question (Bond 685; Herschel ϵ).

Coming now to the region of the nebula about the stars of the trapezium, it will be seen from Fig. 31 that the bright lines are considerably widened where they intersect the spectra of the trapezium stars. In this case the hydrogen line at λ 4340 is widened very little on the less refrangible side, while, on the more refrangible side, the widening is nearly as great as its own breadth. Further, on each side of the line there is a decided break in the continuous spectrum of the stars, giving the appearance of a broad absorption band, with the bright hydrogen line running through it. This appearance is almost exactly reproduced at H_δ.

Dr. Draper[1] appears to have noticed a peculiarity in the hydrogen lines where they crossed the spectra of the trapezium stars in his photographs of 1882. He says:—

"The hydrogen line near G, wave-length 4340, is strong and sharply defined; that at *h*, wave-length 4101, is more delicate; and there are faint traces of other lines in the violet. Among these lines there is one point of difference, especially well shown in a photograph where the slit was placed in a north and south direction across the trapezium; the H_γ line, λ 4340, is of the same length as the slit, and, where it intersects the spectrum of the trapezium stars, a duplication of effect is noticed. If this is not due to flickering motion in the atmosphere, it would indicate that hydrogen gas was present even between the eye and the trapezium.

"I think the same is true of the H_δ line, λ 4101."

---

[1] *Amer. Journ. of Science*, (3), vol. xxiii, p. 339.

The line at 500 is only feebly impressed in the neighbourhood of the trapezium stars, and no reversal is visible.

It is clear, therefore, that the spectrum of the nebula varies very considerably in different regions. In other words its chemistry varies in different regions.

The General Chemistry of the Nebula.

For this of course we are dependent upon the measurement of the wave-lengths of the lines photographed, and in this measurement it is impossible to give them with an accuracy greater than that expressed in tenth-metres, that is ten millionths of a millimetre.

All the lines thus measured are shown in the table which follows. In all, fifty-four lines have been recorded, and, of these, about twenty are seen without difficulty. The remainder require a favourable light, but no line has been inserted in the table which has not been measured several times by two observers. The spectrum extends from the ultra-violet to the green, and the intensities of the lines on the photographs naturally do not correspond to the visual ones; the F line, for instance, appears stronger than the brightest line in the visible spectrum at λ 5006. The photographic intensities are recorded in the table, six representing the strongest and one the feeblest line.

The Origins of the Lines.

It will be seen from the table, that hydrogen enters largely into the composition of the gases of the nebula. H , H_γ, H_δ, H_ϵ, and the ultra-violet series, certainly as far as H_κ (new notation),[1] are all present.

It is worthy of remark, however, that while, as previously

[1] Vogel, *Ast. Nach.*, 3198, 1893.

Table I.—Lines Photographed in the Spectrum of the Orion Nebula, February 10, 1890.

Wave-length.	Photo-graphic intensity.	Probable origins.	Wave-length of probable origins.	Remarks.
3707·4	2			
3715·4	1			
3729·5	6			
3743·6	1			
3752·6	1	H	3752·05	H_κ
3770·6	1	H	3770·70	H_ι
3796·6	2	H	3798·00	H_θ
3833·6	2	H	3835·60	H_η
3847·6	1			
3855·7	1			
3868·7	4	He	3867·61	
3887·7	4	H He	{ 3888·79 3889·15	He H_ζ
3902·7	2			
3910·7	1			
3933·7	2	Ca	3933·83	K line, Solar Spect.
3941·7	1			
3949·7	1			
3968·7	5	Ca H	{ 3968·63 3970·25	Ca H_ϵ
3984·7	1			
4000·7	3			
4010·7	2	Gas X	4009·42	
4025·7	3	He	4026·34	
4041·6	1			
4054·6	2			
4067·7	2			
4086·7	1			
4101·8	6	H	4101·85	H_δ
4120·8	1	He	4120·97	
4129·8	1			
4142·7	1	Gas X	4143·92	
4154·7	2			
4167·7	1	Gas X	4169·13	
4204·6	1			
4226·6	1	Ca	4226·90	Flame line
4234·6	1			
4269·6	2			
4340·6	6	H	4340·66	H_γ
4385·7	1	Fe	4383·72	Strongest flame line of iron.
4389·7	2	Gas X	4388·10	
4410·7	1			
4426·7	2			
4471·8	4	He	4471·65	Lorenzoni's *f*.
4495·8	3			
4539·8	2			
4627·8	2			

TABLE I—*continued.*

Wave-length.	Photo-graphic intensity.	Probable origin.	Wave-length of probable origins.	Remarks.
4715·9	2	He	4713·25	
4735·9	1	C	4737·18	
4825·0	3			
4840·0	2			
4862·0	6	H	4861·49	H_β
4897·9	3			
4923·9	3	Gas X	4922·10	
4958·0	4			
5007·3	5	Mg	5007·18	"Chief" line

stated H_γ is the strongest line in the whole spectrum, and H , H_δ, and H_ϵ are also strong, the ultra-violet hydrogen lines are amongst the weakest.

Next to H_γ, the line λ 373 is the most intense. In 1887 I suggested that this line was one of the members of the triplet seen in the spectrum of burning magnesium. The wave-length could not be finally determined, but it was probably near λ 3729.

This value, however, will require correction for motion in the line of sight. If Mr. Keeler's values[1] for the motion be accepted, and the earth's orbital velocity be allowed for, the correction will be about 0·22 tenth metre towards the red. This will bring the nebular line slightly nearer the least refrangible member of the magnesium triplet. Further measures of photographs taken with higher dispersion are necessary in order to settle this point.

The lines next in importance to those already mentioned, are near wave-lengths 4471, 4495, and 3868. The first of these, the strongest between H_β and H , is probably the line observed by Dr. Copeland in 1886. With reference to this line, I wrote as follows in 1889.[2]

[1] *Proc. Roy. Soc.*, vol. xlix, 1891, p. 400.
[2] *Ibid.*, vol. xlvii, p. 30.

"The observations of Dr. Copeland have now, I think, established the identity of the yellow line, in the nebula of Orion at all events, with D^3. In a letter to Dr. Copeland I suggested that the line at λ 447 was, in all probability, Lorenzoni's *f* of the chromosphere spectrum, seeing that it was associated both in the nebula and chromosphere with hydrogen and D^3. This he believes to be very probable. The line makes its appearance in the chromosphere spectrum about 75 times to 100 appearances of D^3, or the lines of hydrogen."

I have recently discovered that the above line is due to the cleveite gases, and I have found that all the lines of these gases in the region covered by the photograph are present, so that the existence of these gases in the nebula is as clear as is that of hydrogen (see *ante*, p. 39).

Only a small proportion of the lines now mapped can be ascribed to metallic origins, but these, it will be seen, are among the chief lines in the spectra of the elements concerned. The table shows that a number of the lines still appear to have no terrestrial equivalent, but they are present in the spectra of other celestial bodies.

THE PLANETARY NEBULÆ.

We can pass from such a nebula as that of Orion to the well-known planetary nebulæ. Almost all the knowledge which we have of these nebulæ we owe to the labours of Sir William and Sir John Herschel. So far as appearance goes, we have in these planetary nebulæ almost to deal with a planet like Jupiter, except that we do not see the belts. That is why these bodies are called planetary nebulæ; they give us the idea that we are dealing with discs. Among them we pass from a nebula which is simply discoidal to others in which we find a very faint disc, including a much brighter condensation at the centre. There are others in which we get a very strong condensation towards the centre; there is a very considerable difference in the intensity of the light given out as the centre is approached.

We owe most of our present knowledge of the spectra of the

planetary nebulæ to the labours of Professor Campbell of the Lick Observatory, and to Professor Pickering of Harvard. The result of their work goes to show a great similarity between the spectra of these bodies and that of the nebula of Orion.

The Chief Nebular Line.

Having given an idea of the new researches, I next come to consider the only point which has given rise to discussion. This has to do with the origin I assigned to the chief nebula line, the wave-length of which I took from the old observations I had brought together. While Dr. Huggins had declared for nitrogen in some unknown form, certain observations, fully recorded in the *Meteoritic Hypothesis,* had suggested to me that the line might owe its origin to magnesium.

I have already shown how Dr. Huggins in his address as President of the British Association had, apparently from quite independent inquiry, announced my main contention, namely, that there is an evolution of celestial forms, and that nebulæ and stars do belong to the same order of celestial bodies, and had withdrawn his earlier statement in the opposite sense as having been made partly on theological grounds.

I am now compelled to add, but I wish to make my statement with the utmost courtesy, that a complete study of the literature shows that he was quite familiar with my work all the time, and that while he thought fit to republish my main contention as his own on the one hand; on the other he was engaged in attempting to throw discredit on my work, and to conceal his retreat after the manner of the sepia by a great cloud of ink—printer's ink, referring to a minor point.

It will be clear to most people that the moment the main contention of a discussion has been accepted it is not necessary to discuss the arguments, especially the least important one, which have led up to it, but this was Dr. Huggins' position.

I have stated in the *Meteoritic Hypothesis* the considerable amount of research which was undertaken to ascertain the true wave-length of the nebular line, and to account for the various curious appearances which had been noticed in connection with it; some of these I have again referred to in the present chapter, p. 91.

As a result of all this inquiry I stated that its position in the spectrum did *not* correspond with the line of nitrogen, to which element Dr. Huggins ascribed it; that with the highest dispersion I could employ it *did* correspond with a fluting of magnesium, and that its fluted nature would account for the strange appearances to which reference has been made.

One word before I pass on, regarding the difference between line and fluted spectra. Line spectra will be familiar to all my readers; they are built up of the images of the slit irregularly spaced throughout the whole length of the spectrum. The two spectra of barium and iron, given in Fig. 33, may be held to represent this type.

1
2

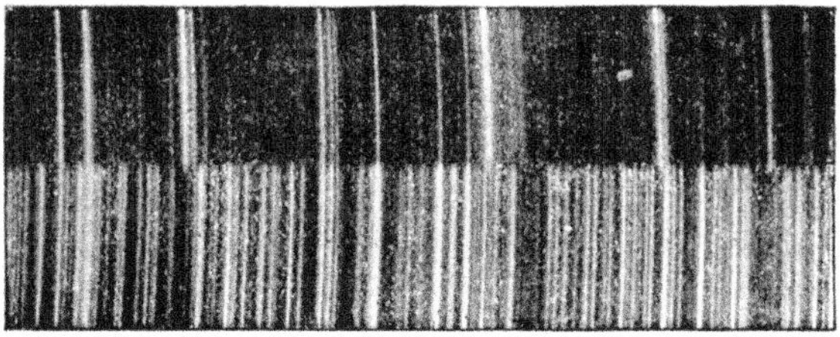

FIG. 33.—A small part of the Line spectra of barium and iron. 1, Barium. 2, Iron.

Besides what we term line spectra, there is another thoroughly well-recognised class, which we call fluted spectra, because it reminds one of the flutings of a column. In these flutings, instead of the lines being distributed irregularly, as in the case of iron and barium, we get a beautiful rhythm from one part where the light rapidly degrades to another where there is an enhancement of the light, followed by another degradation, and so on. Indeed, we not only get main flutings, but we get subsidiary flutings.

One of the flutings seen in the spectrum of carbon is illustrated in Fig. 34.

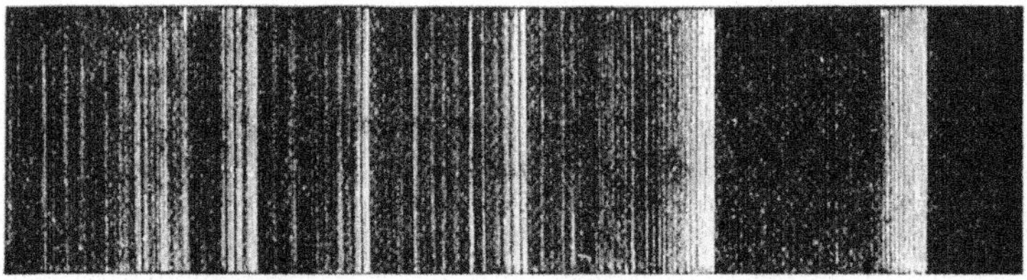

FIG. 34.—One of the flutings of carbon.

The difference in the appearance of spectral lines and flutings having been explained, I now go on to state that the luminosity referred to, as seen in the spectrum of magnesium in my experiments at the place of the chief nebular line was a well developed *fluting*, and not a *line*. I give a photograph of it in Fig. 35.

I have already said that with the greatest dispersion I could employ, the place in the spectrum of this fluting agreed with that of the chief nebular line.

That, then, was one argument out of a great number in favour of the view that the luminosity to which the bright line of the nebula was due might really be produced in the nebula by magnesium vapour at a low temperature.

Next, an additional argument for that view was found in the fact that almost every observer, including Dr. Huggins himself, had stated that, as seen in the spectrum of a nebula, the line did look somewhat different on one side to what it did on the other, and references were made to its being more degraded on the blue side. I had frequently observed myself that the line was degraded to the blue, never to the red, over parts of the nebula

1

2

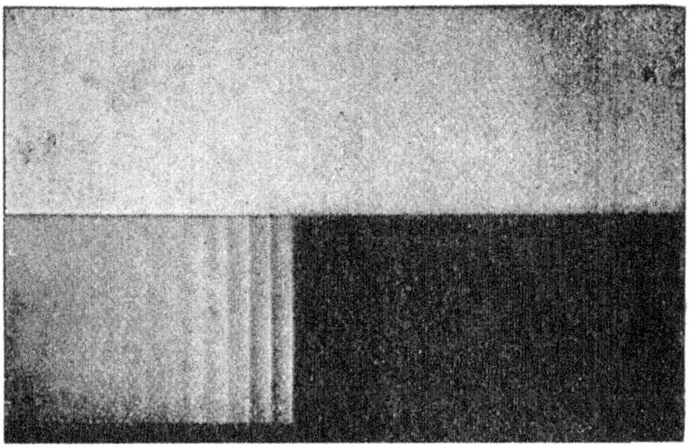

FIG. 35.—Spectra of burning magnesium compared with solar spectrum. (1) Sun. (2) Magnesium.

of Orion which were more brilliant than the others: and at the same time that another line, visible at the same time, instead of being degraded to the blue like this one, was equally eased off on both sides (see *ante*, p. 91), so that the argument was complete that the appearances presented by the line were not due to any instrumental defect, because, in that case, all the lines would have behaved in the same abnormal manner. Hence, then, I found myself justified in concluding and subsequently stating (1) that the position of the magnesium fluting was coincident with the line of the nebulæ in the apparatus which I used, and (2) that it resembled it in appearance.

The question was one of such extreme interest that I spared no pains in increasing the dispersion employed, and no care in trying to eliminate all instrumental errors. Even a siderostat and a firmly supported horizontal telescope were employed to get rid of flexure.

In the end a very much more powerful dispersion than had been employed by Dr. Huggins in his first observations, and much more powerful than had been employed by myself in my first investigation, was employed. It is as well, however, to state that what I wished to do in the first investigations was to

FIG. 36.—Showing object glass of horizontal telescope used with siderostat.

understand and to clearly follow the observations which had been made previously by others; if, therefore, I had attempted to go over the ground in the first instance with instruments ten times more powerful, giving me ten times finer results than my predecessors had obtained, it would have been the worst possible way to go to work, because it was essential for me to make the necessary comparisons with the old observations while not exceeding the instrumental means which had been employed to obtain them.

The long and short of my various methods of observation was that they seemed entirely to confirm the idea which I got in the first instance from using telescopes and spectroscopes of very much smaller power. That, however, fortunately for science, did not satisfy Dr. Huggins; he very wisely appealed to the American astronomers, and I am glad to say that the skilful observers of the Lick Observatory took up this work with interest, and employed instruments in the investigation more powerful than any I possessed, thus carrying matters a stage further. There were really two distinct bits of work to be done; first of all, one wanted the exact position of the line in the nebulæ, and after having got its right position, its origin could be thought out. We wanted also to see what the real physical appearance of the line was, *i.e.*, whether it was most likely a line or a fluting. It is not a little curious to note that all the statements which had been made suggesting a fluted character of the line were at once withdrawn when I referred its origin to the magnesium fluting.

The Lick telescope is one of very considerable power indeed, and it is so solidly built that a very powerful spectroscope can be put on one end of it and used under almost the best possible conditions for determining the position of lines. Still the Lick telescope is not the best possible telescope to employ for any branch of work connected with nebulæ, if the work requires a great amount of light, because the longer the telescope, the larger the image which the object-glass gives; for instance, if we are dealing with a nebula one degree in diameter, if our one degree is written on a circle with a radius of 60 feet, it will be a very much bigger thing than if on a radius of 10 feet, so we get a large image without increasing the light, and therefore are spreading the light over a very large area. As the slit of the spectroscope is a very small thing, all the light which is thrown outside the slit is of no use for our spectroscopic observation, so, whatever the size of the spectroscope may be, we want to deal with the smallest and brightest possible image in order to get the best use out of the spectroscope, and that cannot be done with a long focus telescope. However, the important question for the American observers was to determine the exact position of the line, and Mr. Keeler soon obtained some very interesting results.

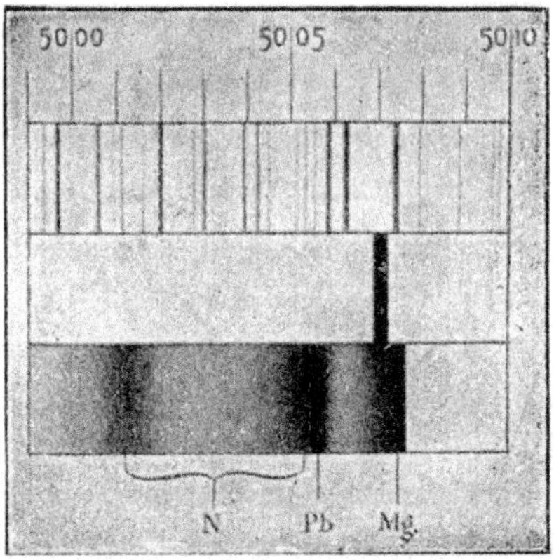

FIG. 37.—Normal position of the chief nebular line, according to Keeler.

Fig. 37 represents the way in which Dr. Keeler, following my example (see *Meteoritic Hypothesis*, p. 301), puts his last result. The upper part is a representation of the solar spectrum; the numbers represent wave-lengths on Rowland's scale. According to his latest value the wave-length of the nebular line is 5007·05. He also shows in relation to it the lines of nitrogen as well as the fluting of magnesium, and we see at once that, although according to this drawing the magnesium does not quite correspond with the line of the nebulæ, it is very much nearer to it than is either the line of lead or the lines of nitrogen.

The publication of this result necessitated a fresh investigation, to see what the exact facts were when we no longer compared the nebula with magnesium, but compared the magnesium with the solar spectrum, and, therefore, sought the true position in which to place the magnesium line in relation to the solar spectrum.

This gave me the value 5007·18; so that the difference was brought down to a difference of wave-lengths represented by a motion of five miles per second.

Without labouring the matter further, we can say that the more the work done on this question, the more and more coincident have these lines become, and there are some considerations which have not yet been taken into account.

In order to give an idea of the relative accuracy which all these references to wave-length indicate, let us suppose that we are trying to define the position of a place in London on an E. and W. line running through Charing Cross. Assuming Dr. Keeler's value to be absolutely true—and I expect it is as near the truth as we are likely to get for some time—we will suppose that it represents the nebular line situated on the statue of King Charles at Charing Cross. When Mr. Huggins first measured it (1868) he brought it to the East India Docks; his next attempt (1872–74) brought it to Hammersmith. Dr. Keeler's first observation (1889–90) brought it to Albert Gate; his next, in 1891, brought it to St. James's Palace. Subsequent work at Kensington, not yet completed, has brought it nearer still.

I trust I shall not be thought to be exceeding the bounds of decorous criticism when I remark that while Dr. Huggins has referred to the "inaccuracy" of my work in relation to this line, which is, apparently, indicated by Dr. Keeler's results, he has never pointed out the three times greater inaccuracy of his own, for anyone can gather from Dr. Keeler's diagram that the nearest nitrogen line is three times further removed from the position of the nebula line than is the magnesium fluting.

Nobody believes in the nitrogen constituent of the nebulæ now; and I presume Dr. Huggins has withdrawn, in fact, if not in words, his statement concerning the coincidence; for, in his address as President of the British Association, in which, as I have already stated, he withdrew his published statement as to the position of the nebulæ among the various bodies that people space, he remarks:—

> "The progress of science has been greatly retarded by resting important conclusions upon the apparent coincidence of single lines in spectroscopes of very small resolving power."

An *apologia* of which everyone will see the propriety.

I have referred to this point somewhat at length, although the coincidence with the fluting of magnesium is not fundamental for the establishment of the view which even Dr. Huggins has now accepted, in the way I have already stated.

Many who are, or rather were, in favour of the view that the temperature of nebulæ is very high, while giving up Dr. Huggins' view of an unknown form of nitrogen, have objected to my assigning the chief line to magnesium on the ground that it is a low temperature line, and have suggested that it is a line sometimes seen in the solar chromosphere. It may be worth while to remark in this place, therefore, that the cleveite gas which accounts for so many lines in the solar chromosphere, has no line near the place.

CHAPTER VIII.—THE NATURE OF NEBULÆ.

RICH STORE OF NEW FACTS.

IN the previous chapter the chemistry of the nebulæ has been dealt with in some detail. It has become abundantly clear that with the facts before us, now garnered in considerable quantity, it is impossible to regard them merely as *residua*, as masses of nitrogen and hydrogen. The new wealth of observations has not only thrown a flood of new light on the origin of the spectrum, but has enormously increased the complexity of it.

I propose in this present chapter, having cleared the chemical ground, to see what the probable nature of the nebulæ is to provide them with such a chemistry. This discussion will enable us to supply a test between the two rival hypotheses, one that the light of the nebulæ is due merely to incandescent permanent gases taken at random, the other that it is produced by the collision of meteorites.

In my first attempt to discuss the spectroscopic results, the facts I had at my disposal were very few, sixteen lines only had been recorded (*Meteoritic Hypothesis*, p. 287), whereas now I have obtained fifty-four lines in one photograph! In the first instance, I had to examine into the validity of Professor Tait's suggestion by referring to hydrogen, carbon, and magnesium. With regard to the latter substance I may remind my readers that, after studying all the nebula observations which I could lay my hands on, I went on to observe and to experiment myself. Meteoritic dust was scattered along the capillary in a horizontal spectrum tube connected with a Sprengel air-pump, so that an electric spark could be passed through the dust; an image of the end-on capillary was formed by means of a lens on the slit of the spectroscope, and an arrangement for an ordinary electric spark in air served for a comparison spectrum. The point was to see whether there was any probability that Professor Tait's suggestion was right, by examining the spectrum of meteoritic dust.

What I found was that in the spectrum of dust from several meteorites so examined, there was a line very near the position which had been stated by Dr. Huggins to represent the actual position of the chief line seen in the spectrum of the nebulæ. That line I was able to trace home by comparative work, not to nitrogen, but to olivine, a substance which occurs in almost all meteorites even in iron meteorites; and not only to olivine, but to one of the constituents of it, which is magnesium.

But magnesium now is not the only metallic substance we can consider, for, in the previous chapter, I have shown that calcium, iron, and other metallic lines are now at our disposal.

It has long been known that all meteorites give off hydrogen and carbon compounds when heated, and Professor Ramsay states that he has recently detected cleveite gas given off in the same way.[1] It is equally well known that iron, calcium, and magnesium are among the most important metallic constituents.

At the outset, therefore, an important advance was made, since on the new hypothesis, the spectrum of the nebulæ must be built up by lines and flutings of the gases and vapours given off by meteorites when heated by collisions.

The Conditions to be Studied.

The first point in the discussion, therefore, is to consider not what is happening in the nebulæ, but what must happen in

---

[1] I have not succeeded in this.

a laboratory when meteorites and the substances they contain are exposed to various temperatures; this will enable us to consider what will happen to them when they are heated by collisions.

Now, obviously the first things to be given off by meteorites heated by collisions, and to remain visible longest, are the *permanent gases* which they contain; metallic vapours then produced would, of course, soon condense and become invisible on this ground alone. Hence we should expect hydrogen, the cleveite gases, and carbon compounds to be represented in the spectrum.

This will be true for collisions generally provided the temperature produced by them is sufficient to compel the meteorites to give out their "occluded" gases. When once out they may remain incandescent by virtue of their temperature, or even electric excitation may be postulated to render or keep them luminous.

If one grant electrical excitation, a temperature low enough to expel the occluded gases may be sufficient to produce the luminosity of the permanent gases.

But to study which metallic lines we are to expect, we must particularise somewhat; there must be under the conditions assumed collisions and collisions. In the laboratory, flame, arc, and spark spectra of metallic substances are recognised in the above order of increasing temperature, and it is obvious that collisions, from grazes only to end-on encounters, with the different heat conditions, must give us different spectra.

This premised, let us next pass to the conditions present in a nebula on the meteoritic hypothesis.

In order to make the distinction perfectly clear between the metallic lines in the possible sources of nebula spectrum let me ask the reader for one moment to conceive himself in the middle of the gigantic battle which is going on, if the hypothesis be true, between meteoritic particles in such a nebula as that of

Orion. We have particles rushing together in all possible directions, particles, no doubt, different in origin. We shall expect, among those millions and billions and trillions of collisions, to get a very considerable number of grazes; and the whole point of collisions among physical particles is that, if two things go straight at each other, we get an end-on collision, which may be bad for one or both of the bodies concerned; the temperature under these circumstances is at a maximum. But the number of grazes, or near misses, must be very much greater than the number of end-on collisions; in such a case as we are imagining, there will be an immense number of grazes. What will a graze do? It is simply a slight collision; the temperature developed by it will be small; we shall, therefore, get the production of vapours at a low temperature, and if we get any luminous effect at all, it will be one proper to the vapours at low temperature. So that on first principles we should expect in such a nebula as the one we are discussing to get a very large number of grazes, giving us low temperature effects, and a very much smaller number of end-on collisions, giving us very high temperature effects, if the number of collisions due to this cause is sufficient to enable us to detect them.

To summarise: We are justified in assuming that the most numerous collisions will be partial ones—grazes—sufficient only to produce comparatively slight rises in temperature. The metallic lines visible in the nebula spectrum, so far as it is produced by this cause, will, therefore, depend upon the phenomena produced in greatest number, and we may hence expect to find the low temperature lines of various metallic substances.

In addition to the large number of partial collisions there may be a relatively small number of end-on collisions, producing very high temperature, and, so far as this cause is concerned, there may be some metallic lines produced which are associated with very high temperatures.

Hence the bright lines should have three origins, namely:—

(1) Non-condensable gases driven out of the meteorites by collisions generally.

(2) Low temperature vapours produced by a large number of feeble collisions.

(3) High temperature vapours produced by a small number of end-on collisions.

Now, what is the truest way of determining the results of high temperature in a body such as the nebula of Orion? This is what I said on this point in 1894,[1] and I have nothing to withdraw:—

"We want to know what the possible results of the highest temperature will be. The natural thing, I think, is to go to the sun, which is pretty hot, and then find out the very hottest place, which we can do by means of our spectroscopes, and then study very carefully, for years even, the spectroscopic indications in that particularly hottest place of the nearest star which we can get at. I hope you will acknowledge that that is a philosophic way of going to work. Thus we are landed in what is called the chromosphere of the sun. The upper atmosphere of the sun must be cool, but the chromosphere is a thin envelope some 5,000 or 10,000 miles thick, just outside the photosphere, agreed to be the hottest part of the sun within our ken, and therefore any metallic lines which we see special to that region are called chromospheric lines, and they should be proper to high temperatures."

Recent research has shown that the most prominent metallic lines in this region, besides those of hydrogen and the cleveite gases, are certain lines seen in the spectrum of calcium, strontium, iron, manganese, and magnesium.

The number of these lines in comparison to the number of Fraunhofer lines is exceedingly restricted, and it may be further added that these lines are all enhanced at the temperature of the most powerful electric spark we can employ in such inquiries.

From the solar point of view then, the sun being a thing that we can get at better than any of the other stars, because it is so

[1] *Nature*, vol. lii, p. 13.

near to us, we are justified in saying that dealing with known substances these lines represent the spectrum of the hottest part of space about which we can be absolutely certain. Hence it is very interesting to inquire whether or not these lines exist in the nebulæ as representing high temperature.

STUDY OF THE NEW FACTS.

Such, then, are the phenomena we have to look for. Next let us examine into the facts the new work has brought us.

I will begin with (1) the non-condensable gases driven out of the meteorites, that is, the lines of those substances which occupy the greatest volume (or largest area in a section); in other words, the lines of those substances which are driven *furthest* out from the meteorites and occupy the interspaces, where possibly they may be rendered luminous by electricity. Chief among these, from laboratory experiments, we should expect hydrogen; next, from the same experiments, we should expect gaseous compounds of carbon, and finally, as the result of the latest work, helium and its associated gases.

The table on pages 95–6 shows that in the nebula of Orion we get all the first and last in abundance. In this nebula the traces of the compounds of carbon are not so obvious, but, on the other hand, in the nebula of Andromeda, we get them without hydrogen and helium. Taking both nebulæ we get all three then fully represented.

It is important to point out here that when, in 1888, the close connection between comets and nebulæ was forced upon me by the discussion of the observations, the spectrum of carbon had not been announced in the nebulæ, but I predicted that, if the nebulæ were carefully observed, we should find in them sooner or later indications of it for the reason that in almost every comet which has been observed, the spectrum of carbon, or of some compound of carbon, is the strongest and most obvious feature which is presented to us. In 1889, *i.e.*, only the next year, matters were made very much clearer by the discovery, by Mr. Fowler and Mr. Taylor, of the spectrum of carbon in the nebula of Andromeda, so there was a prediction verified, and such verification is always a very precious test, since it helps us to know whether one is going right or wrong. The discovery of carbon undoubtedly strengthened the hypothesis very much. Observers are not all of one opinion as to whether the so-called "carbon spectrum" so marked a feature in comets is due to carbon or a compound of carbon, but this is immaterial for that part of the hypothesis now under discussion.

Next a word with regard to the appearance of the cleveite gases. In regard to them we are not yet in the open, too little is known about them. If both they and the hydrogen and carbon compounds are merely occluded in the stones of the supposed meteorites, they may be driven out by relatively low temperatures, and if they glow in the nebulæ, by electricity, we have no arguments in favour of a transcendental temperature.

If, however, we regard another possibility that hydrogen and helium represent primordial states of matter, then undoubtedly their presence in the nebulæ might indicate a very high temperature.

There are some points in favour of this view. Hydrogen and the cleveite gases do not extend high in the solar atmosphere, as hydrogen was once thought to do.

How comes it that these so-called permanent gases cannot be traced above the sun's chromosphere? Do they enter into combination at a slight reduction of temperature, and if so, is not their appearance both in sun and nebula an indication of high temperature after all?

There is another matter, as I pointed out in the *Meteoritic Hypothesis* (p. 388); one of the lines of hydrogen (H), in the spectrum of *o* Ceti, seems absent from the spectrum, as if the special radiation was absorbed by the calcium vapour of the same period; if this be so then in that star, which I regard as an advanced nebula, the most vividly incandescent hydrogen is being produced from the meteorites, and does not exist exclusively in the interspaces.

I next come to (2) the metallic lines due to the results of grazes.

These are represented in the nebula of Orion by flame lines of iron and calcium; of these calcium is the most secure, as it has long been known that the line at 4226 is only seen strongly developed at low temperatures, while the lines at H and K are only strongly developed at high temperatures. In the solar chromosphere the intensities of H and K to 4226 are as 10 to 1, in the nebula of Orion they are only as 2 to 1.

There is also to be mentioned the possible representation of the chief magnesium fluting in the green by the chief nebula line at 5007; that fluting of magnesium is the lowest temperature indication of the existence of the metal. If magnesium becomes luminous at all by virtue of its temperature, one of the first things revealed to us spectroscopically is the fluting in question.

Finally we have to deal with (3) the metallic lines, the results of end-on collisions. It is here that the evidence is almost wanting. We do not, as I have shown, get the enormous development of H and K seen in the solar chromosphere, neither can the high temperature lines of magnesium or strontium be traced.

The remarkable result of the inquiry is then to show that the requirements of the hypothesis with regard to nebulæ are met in every point so far considered by the new facts, on the assumption that the temperature is not too high.

That is we certainly find in the spectrum of the nebula of Orion, when it is carefully studied, indications of the gases which are known to be occluded in meteorites, and which are perfectly prepared to come out of them the moment we give them the least chance. Then, also, there is the indication of the results of an infinitely great number of grazes in the shape of lines of metals which we see at the temperature of the oxy-hydrogen flame, but which we do not see so well, and alone, at the temperature of the arc and the spark.

The tenour of the new evidence bearing on the view that we have to deal with luminosity produced by collisions in a sparse meteoric swarm is, I claim, all in my favour.

If we were dealing with incandescent gas, the incandescent gas ought to leave off suddenly; but all round the nebula, as it appears in an ordinary telescope, where there appears to be nothing at all, long exposures bring before us, as we have seen in Fig. 29, other portions just as rich in details, just as exquisite in their variety and tone as those ordinarily seen with the naked eye. Such a condition as that cannot be brought about by a mere homogeneous mass of gas at high temperature, but we can explain it quite easily by assuming that in such a nebula as that we are dealing with, the luminosity is brought abou by disturbances, these disturbances giving rise to collisions among the particles which are apt to collide and give out luminosity. The nearer they are to the centre of gravity of the swarm, the greater will be their chance of collision, and the greater will be the luminosity in the central portions.

All the arguments tend to show, therefore, that we are not dealing with gas, but with masses of matter in certain regions of which, in consequence of general action, there is greater luminosity given off by the particles of which the nebulæ are composed; in other regions where there is less action, we have lower temperature and less light.

It is only right that I should state that when my book was written I had not sufficiently thought out the collision conditions, which it now turns out supply us with such a rigid test to apply to the hypothesis.

When the series of lines which were supposed to be associated with high temperatures was first recorded in the spectrum of the nebulæ, I stated that possibly this might be due to the fact that in regions of space where the pressure may be held to be extremely low, we might be in the presence of chemical forms unfamiliar to us here, because all we know of them spectroscopically here is the result of considerable temperature, and not very low pressure. It was therefore suggested that these lines might represent to us the action of unfamiliar conditions in space. If we have a compound chemical substance, and increase its temperature sufficiently, it is dissociated; but imagine a condition of things in which we have that same chemical substance for a long time exposed to the lowest possible pressure. Is it possible that that substance will ever get pulled to bits? If so, we may imagine parts of space which will contain these substances pulled to bits which really constitute finer forms of matter than the chemical substances with which we are generally familiar. So that we may possibly expect to get the finest possible molecules as distinct entities in the regions where the pressure is lowest. These forms are, of course, those we should expect to be produced by a very high temperature brought on by end-on collisions; hence the line of thought is not greatly changed in both the explanations.

It must not be forgotten that the mystery is by no means all lifted from the region of thought brought before us by these considerations. It is perfectly true, as we have seen, that both hydrogen and the cleveïte gases may be made to glow at relatively low temperatures, and, therefore, since they can exist as gases no argument as to the temperature of the nebulæ can be deduced from the appearance of their lines in the nebular spectrum.

I may next remark that a physical structure of the nebulæ, other than that indicated, would be impossible when once it is conceded that they represent early evolutionary forms.

Astronomers, since the time of Rutherfurd, who was the first to deal with stellar classification, have established many different classes of stars as defined by the chemical substances of which their atmospheres seem to be composed, so far as spectroscopic observations enable us to determine their composition. One group of stars is remarkable for the presence of hydrogen in enormous quantities; we assume that because the lines of hydrogen are inordinately thick. In another we get not so much hydrogen, although it is still there; but the predominant substances are iron and calcium. In other stars we get little hydrogen, if any, but carbon in enormous quantities; and, again, there are other substances, the quantities of which vary enormously, I refer to helium and its associated gases. Now, if stars contain all these different substances, and if they represent epochs of evolution—and this point is now conceded—they must be produced from something which actually or potentially contained these substances; so that there, again, we get an important argument in favour of the chemical complexity of the nebulæ.

The more inquiry proceeds the more does the strictest chemical similarity between all classes of heavenly bodies make itself manifest.

The presence of carbon, to take an instance, first recognised in comets is now recognised both in nebulæ and stars. In some stars carbon exists in enormous quantities. Here, then, we are in the presence of the fact that the statement that there is an enormous chemical difference between nebulæ and stars, is shown spectroscopically to be unfounded, while the evidence also goes to show that there is a close connection between nebulæ and comets.

I have already stated that authorities are not all of the same opinion with regard to carbon; some hold that the spectrum is produced by carbon vapour, others that it is due to carbon compounds; but the more it is held that it is really due to a compound, the lower the temperature of nebulæ must be.

Result of the Discussion.

The total result of all this inquiry has been to justify the statement I made in 1887, that the mean temperature of the phenomena brought before us in nebulæ is distinctly low. This result is of extreme interest and importance, because, remembering what was said about the objection to Laplace's view of high temperature gas because it violated the laws of thermodynamics, we have now, after minute study, come to a conclusion regarding the structure of these nebulæ, which is quite in harmony with the laws of thermodynamics, and the new view is making way. We hear no more of a finer form of nitrogen, and even Dr. Huggins has now come to the conclusion that in nebulæ we have distinctly a relatively low temperature. His change of front on this point is as remarkable as on that touching the constitution of the nebulæ; and, again, I can claim it as a valuable, independent confirmation of my views, for he makes no reference to them. Even in 1889 Mr. Huggins held that "They [the nebulæ] consist probably of gas at a high temperature;" but in the address of 1891, to which I have already had occasion to refer, he gives this view up, and refers to "the much lower mean temperature of the gaseous mass *which we should expect at so early a stage of condensation.*"[1]

---

[1] "On account of the large extent of the nebulæ, a comparatively small number of luminous molecules or atoms would probably be sufficient to make the nebulæ as bright as they appear to us. On such an assumption the average temperature may be low, but the individual particles, which by their encounters are luminous, must have motions corresponding to a very high temperature, and in this sense be extremely hot.

"In such diffuse masses, from the great mean length of free path, the encounters would be rare but correspondingly violent, and tend to bring about vibrations of comparatively short period, as appears to be the case if we may judge by the great relative brightness of the more refrangible lines of the nebular spectrum.

"Such a view may perhaps reconcile the high temperature which the nebular spectrum undoubtedly suggests with the much lower mean temperature of the gaseous mass, which we should expect at so early a stage of condensation, unless

I am also glad to say that Dr. Keeler is also perfectly prepared to accept the view I have been insisting on.[1] So that, if the opinion of astronomers of repute is worth anything, we do seem to have arrived at very solid ground indeed on this point, so far as as a consensus of opinion can make any ground solid.

It is, then, generally conceded that the first stage in the development of cosmical bodies is not a hot gas, but a swarm of cold meteorites. From the point of view of evolution, keeping well in touch with the laws of thermodynamics, the nebulæ must begin cool if they are to develop into hot stars.

To sum up, it may be said that the new work strengthens the hypothesis in showing that, whether the spectroscopic phenomena of the nebulæ are produced by the collisions of cool meteorites or not, cool meteorites, under the conditions postulated, would certainly produce them.

---

we assume a very enormous mass; or that the matter coming together had previously considerable motion, or considerable molecular agitation."—"Huggins Address," Cardiff, 1891, p. 21.

[1] *Pub. Lick Observatory*, vol. iii, 1894, p. 225.

CHAPTER IX.—MANY SO-CALLED STARS ARE NEBULÆ.

The Bright-line Stars.

We come now to the third new point of view. Many apparent stars, which have bright lines in their spectra, are really centres of nebulæ, *i.e.*, of meteoritic swarms.

The association of the nebulæ with the bright-line stars was, I think, first suggested by me in 1887.[1]

In the above very simple statement we have perhaps the very greatest and the most fundamental change which has been suggested by the new hypothesis. I am quite certain that all conversant with text-books of astronomy will be perfectly familiar with the statement that all stars are distant suns. I have written that myself several times, but I now know that it is not true. Some stars, instead of being distant suns like our sun, a condensed mass of gas with a crust gradually forming on it and a thick atmosphere over it, are simply the brighter central condensations of nebulæ, whether they be like that of Andromeda, or planetary nebulæ, or such a nebula as that of Orion. The idea is perfectly new and completely different from the old one, which taught that all stars were suns.

Associated with this view we have the statement that stars with bright lines are closely associated wtth nebulæ, as evi-

---

[1] *Proc. Roy. Soc.*, vol. xliii, p. 144.

denced by their spectra. There is one method which enables us to compare the bright lines in stars like γ Cassiopeiæ with the nebulæ, as it gives us an opportunity of determining whether or not the bright lines seen in the so-called bright-line stars are or are not the same as the bright lines seen in nebulæ. In my first inquiry[1] in this direction, which consisted of a statistical statement of the number of times certain lines were seen in the spectra both of nebulæ and of bright-line stars, the observations were so few at that time that only nine lines were found to be coincident.

There were seven in the nebulæ not in the stars, and six in the stars not in the nebulæ, only twenty-two lines in all to deal with!

The harvest of facts is much more abundant now.

So far as the planetary nebulæ are concerned, this grouping has been abundantly confirmed by Professor Pickering's work on the bright-line stars, and by the visual observations of Professor Keeler.

Professor Pickering, who is one of our very highest authorities in all these matters, having tabulated a much larger number of lines, accepted at once the grouping together of stars having bright lines in their spectra with the nebulæ.

He wrote in 1891[2]:—

"Owing to the similarity of the spectra of the planetary nebulæ and the bright-line stars, they may be conveniently united in a fifth type."

It is clear then, that in this particular, Professor Pickering accepts my view.

Mr. Keeler also confirms the view. He writes[3]:—

"The spectra of the nuclei of the planetary nebulæ have a remarkable resemblance to the Wolf-Rayet and other bright-line stars. . . . The D^3 line appears in the central condensation of a number of bright nebulæ,

---

[1] *Proc. Roy. Soc.*, vol. xlvii, p. 38.
[2] *Ast. Nach.*, 3025, 1891.
[3] *Proc. Ast. Soc. Pacific*, vol. ii, No. 11, November 29, 1890.

and with sufficient light would probably be seen in many of them, and this line is also predominant in most of the bright-line stars."

This confirmatory evidence, it will be seen, deals alone with the planetary nebulæ. It became necessary, in order to be able to apply further tests, to study the relationship with other nebulæ as well, and that is why while Pickering and Keeler were working in America on the planetary nebulæ, I determined to attack the nebula of Orion.

I now proceed to show that the relationship indicated between the planetary nebulæ and bright-line stars also holds good for such a nebula as that of Orion.

The bright lines seen in the visual spectra of the two classes of nebulæ have long been known to be identical, and a comparison of the Westgate photographs with the results obtained by Professor Pickering, and the more recent work of Gothard,[1] and of Professor Campbell, at the Lick Observatory, on the spectra of the planetary nebulæ,[2] has shown that the similarity also extends to the photographic region.

The fact that some of the Orion nebula lines were apparently coincident with lines in the bright-line stars, was recognized at an early stage in the reduction of the Westgate photographs, and in the preliminary note I wrote as follows:—

"It is a very striking fact that some of the chief lines are apparently coincident, although the statement is made with reserve, with the chief bright lines in P Cygni, a magnificent photograph of which I owe to the kindness of Professor Pickering; it is one of the Henry Draper Memorial photographs."

The bright lines here referred to were those of hydrogen, and lines at 4025 and 4471. (These lines we now know to be due to the cleveite gases.) All these have since been photographed at Kensington, in the spectrum of P Cygni, and there is no longer any doubt as to their identity with bright

---

[1] *Ast. and Ast. Phys.*, 1893, p. 51.
[2] *Ibid.*, p. 276.

lines in the nebula. Additional bright lines in the spectrum of P Cygni, photographed at Kensington, are also seen in the nebula, as shown in the following table:—

Orion Nebula.	P Cygni (Kensington).
3968·7	3968·7 H_ϵ
	4015·7
4025·7	4025·7
	4035·7
4101·8	4101·8 H_δ
	4147·7
4340·6	4340·6 H_γ
4471·8	4471·8
4715·9	4715·9
	4841·0
4862·0	4862·0 H_β
4923·9	4923·9

The following table shows in a complete form the details of the coincidences of the lines in the spectrum of the Westgate photograph of the Orion nebula with those of planetary nebulæ and bright-line stars, as given by Pickering and Campbell. Only those lines of the nebula which show coincidences are included in this table, but the spectra of the planetary nebulæ and bright-line stars are tabulated in full.[1]

It will be seen that all the lines of the planetary nebulæ, photographed by Pickering, appear in the Orion nebula, while of the twenty lines photographed by Campbell, twelve are present. Of fifteen lines in the spectra of the bright-line stars, eleven appear in the nebula. It must, however, be stated that the dispersion of the Orion nebula photograph is too small to allow of very accurate measurement.

[1] Professor Pickering has been good enough to furnish me with glass copies of his beautiful photographs of the spectra of some of the bright-line stars. The positions of the various lines which he gives are in the main confirmed by the new measures which have been made at Kensington.

Orion nebula. (Rowland's scale.) Intensity Max. = 6	Planetary nebulæ. (Campbell.) (Rowland's scale.)	Planetary nebulæ. (Pickering.)	Bright-line stars, Type I. (Pickering.)	Bright-line stars, Type II. (Pickering.)	Bright-line stars, Type III. (Pickering.)
3868·7 (4)	3867–8				
3887·7 (4)	3888	388	389	389	
3949·7 (1)	—	—	395	—	395
3968·7 (5)	3969	397	398	397	
4025·7 (3)	4026	—	402	402	
4067·7 (2)	4067	—	406	406	407
4101·8 (6)	4102	410	410	410	
4120·8 (1)	—	—	—	—	412
4204·6 (1)	—	—	420	420	421
4340·6 (6)	4341	434	434	434	434
—	4363–4				
4389·7 (2)	4390				
4426·7 (2)	—	—	—	—	443
4471·8 (4)	4472–3	447	—	447	
—	—	—	—	451	451
—	—	—	454	455	455
—	4574				
—	4595				
—	4610				
—	4631–40	—	462	464	464
—	4663				
—	4686–8	—	469	469	
4715·9 (2)	4714–6	470			
—	4743				
4862·0 (6)	4862	486	486	486	
4958·0 (4)	4958				
5007·3 (5)	5007	501			

But even more fortunate than this is the fact that Professor Campbell has more recently made a most important and laborious study of these stars at the Lick Observatory, and has observed all the lines in the spectra of a much greater number of stars than was available when I began the inquiry; his measurements are very much more accurate than any that were then at my disposal. What happens when we come to deal with his results? The thing is a thousand times more convincing than it ever was. When we take Campbell's list, we get very many more coincidences than we had when we dealt with Pickering's. So that, the further we go in this inquiry, the greater is the number of coincidences. In the first inquiry there were nine coincidences observed; now we get nineteen coincidences out of thirty-three,[1] or twenty-two out of thirty-three if one includes the planetary nebulæ, hence the number of coincidences is increased by seven. We are, therefore, justified in saying that the more these phenomena are observed, the more closely associated are they seen to be.

Such is the kind of evidence on which we have been compelled to rely to answer the question: Is there any chemical relationship, and therefore physical relationship, between the bright-line stars and the nebula of Orion? It is obvious that the evidence is very strongly in favour of an affirmative statement.

THE ORIGIN OF THE SPECTRAL DIFFERENCES BETWEEN NEBULÆ AND BRIGHT-LINE STARS.

There are some detailed points which are well worth consideration; for although on the hypothesis that these stars are really condensing nebulæ, the lines seen in the spectra should, in the main, resemble those which appear in nebulæ; they will differ for two reasons:—

(1) Owing to partial condensation of the swarm the hydrogen area will be restricted, and the bright lines of hydrogen will lose their prominence; the volume occupied by the carbon compounds will be relatively increased, and the brightness of the carbon bands will be enhanced.

(2) On account of the increased number of collisions more meteorites will be rendered incandescent, and the continuous spectrum will be brighter than in nebulæ.

Now, what do we find?

(*a*) The hydrogen lines are decidedly less prominent in the bright line stars than in the nebulæ. Indeed they were not

[1] *Ast. and Ast. Phys.*, June, 1894.

recorded at all in the eye observations of γ Argûs, Arg.-Oeltz., 17681, of Wolf and Rayet's second and third stars in Cygnus,[1] but they are shown in Professor Pickering's photographs.

(*b*) In my previous discussion of these bodies[2] I showed that there was evidence of a very considerable amount of carbon radiation in the visible region of the spectrum. Subsequent work and an examination of Professor Pickering's photographs have strengthened this view.

(*c*) There can be no question as to the continuous spectrum being brighter in bright-line stars than in nebulæ.

Another point of difference is that the chief nebular line near λ 5007 is not seen in the spectrum of bright-line stars, and this no doubt is due to the relative absence of feeble collisions as condensation goes on. The relative brightening of this line in the spectra of Nova Cygni and Nova Aurigæ (about which I shall write in a subsequent chapter) as the stars faded away, is sufficient evidence that it is associated with low temperature, and hence it is not surprising to find that it is absent from the spectra of the bright-line stars, which on this hypothesis are hotter than the nebulæ, since they are more condensed.

In the bright-line stars so called we have on the hypothesis to deal with central condensations not perhaps very far advanced beyond the stage of the planetary nebulæ themselves, and in this way the almost identity in the spectra is readily explained; but there are not wanting evidences in the heavens that in some cases the condensation has gone much further, in this case the bright line condition remaining, but it is no longer supreme. We have one case in the Pleiades photographed by Dr. Roberts. The photograph indicates that they are not stars; they are nebulæ. What we see in the case of each so-called "star" is obvious; we see the centre of condensation, and more than that, it is not a simple condensation, but there are stream-lines going

---

[1] *Proc. Roy. Soc.*, vol. xliv, pp. 33–43.
[2] *Ibid.*

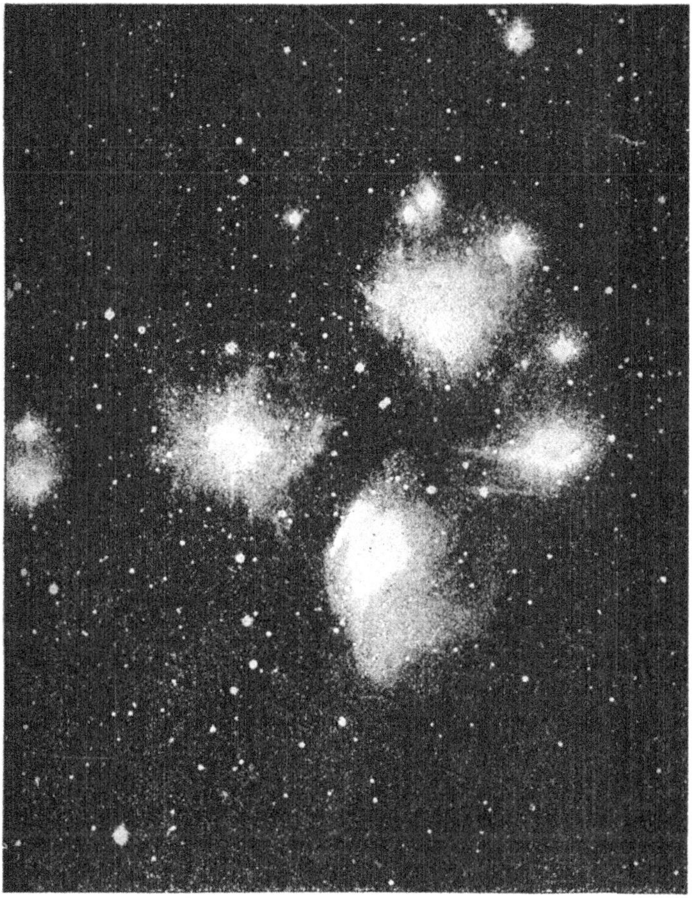

FIG. 38.—The Pleiades (Dr. Roberts).

in all directions, and the maximum luminosity, where we locate the "star," is just at the place where, according to this photograph, the greatest number of streams cut each other, and where, therefore, we should get the greatest possible number of collisions per second of time, and therefore the highest temperature. The main point demonstrated by this photograph, then, is that we are not dealing with stars anything like our sun; we are simply dealing with nebulous condensations. The spectra of the brighter parts of these condensations resemble the spectra of ordinary stars. Broad dark lines of hydrogen are represented in every one; hence, although we are dealing not with a star like the sun, but a meteoritic condensation—a place of intersection of streams of nebulous matter—we get a spectrum such as is generally associated with the spectrum of a star, but the bright-line condition is retained, for bright lines have been seen superposed on the dark ones.

THE EVIDENCE AFFORDED BY THE SPIRAL NEBULÆ.

I propose next to discuss the question whether the long-exposure photographs of Dr. Roberts and others justify or negative the view, founded upon a large mass of spectroscopic evidence, which I put forward in 1887, before any of them had been published.

The view in question was thus stated:—[1]

"The brighter lines in spiral nebulæ, and in those in which a rotation has been set up, are in all probability due to streams of meteorites, with irregular motions out of the main streams in which the collisions would be almost *nil*."

I was careful to state that Professor G. Darwin, when discussing the gaseous hypothesis of Laplace, had already pointed out that

"The great mass of the gas is non-luminous, the luminosity being an evidence of condensation along lines of low velocity, according to a well-known hydrodynamical law. From this point of view the small nebula may be regarded as a luminous diagram of its own stream-lines."[2]

At the time I wrote, in 1887, the nebula in Andromeda was not considered to be a spiral nebula. The most striking representation of it was due to Bond, who drew special attention to two black streaks running nearly parallel to the longer diameter.

---

[1] *Proc. Roy. Soc.*, November 17, 1887, p. 153.
[2] *Nature*, vol. xxxi, p. 25.

It may also be added that in 1887 we knew nothing for certain about its spectrum.

In 1888 Dr. Isaac Roberts published his most admirable long-exposure photographs, which at once established the spiral nature of the nebula; and in the same year the complete discussion of the spectroscopic observations made up to that time led me to predict that if the nebulæ were carefully observed we should find in them, sooner or later, indications of the substance which makes the comet spectrum so very distinct and special. In 1889, that is in the next year, the spectrum of carbon was discovered by Mr. Fowler and Mr. Taylor in the nebula of Andromeda.

I will take the photograph first. The plane of movement in the spiral system is so situated that from our point of space we look at it obliquely; hence the nebula appears elliptical. Still there is no difficulty in seeing that the various streams round the centre of condensation are all of them of a spiral form with certain condensations interspersed here and there along them.

We have a condensation in the prolongation of one of the spirals, and there is considerable clustering of apparent stars along the stream lines. It is important to indicate that we have in these appearances not signs which tell us of the existence of matter merely—so that when we have not the appearances we would be justified in supposing that there was no matter—but an indication of *movement* in matter, so that we may imagine that this nebula and others like it do probably consist of something extending enormously in space beyond the indications which we see, for the reason that near the centre the movements are more violent than they are towards the outside. We are there face to face with the idea that we have to deal with orderly movements. If the movements are orderly, it means that the movements of the constituent particles of the swarm, all of them, or most of them, will be in the same direction; in that case we have the condition of minimum disturbance, and therefore the condition of minimum temperature.

In short, not only have we regular spirals, but in addition to the spiral system there seems to be revealed irregular masses of nebula near both ends of the major axis. Where more than one stream seems to be contending, the brilliancy is enhanced,

Fig. 39.—Nebula of Andromeda, 1887.

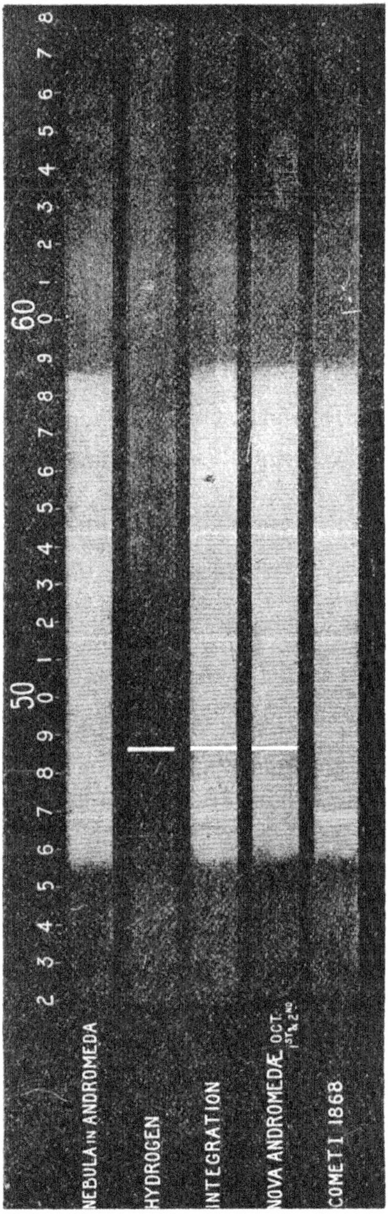

FIG. 40.—Spectra of the nebulæ in Andromeda, Nova Andromedæ, and comet compared. The flutings common to all are those of carbon.

and much irregular luminosity is apparent. On the other hand, in the part of the main nebula most free from these irregularities the spirals are almost invisible.

Next for the spectroscopic observations.

The chief argument urged in favour of the gaseous nature of the nebulæ now is the existence of hydrogen and helium in the planetary nebulæ and in such a nebula as that of Orion; the unknown form of nitrogen has no longer any votaries.

But if the spiral nebulœ be gaseous, why do they not give us the spectra of hydrogen and helium? The spectrum of the nebula of Andromeda is practically the spectrum of a comet, and therefore we are justified in considering it as built up of cometary materials. Now these, as is generally conceded, are meteoritic in their nature.

But this is not all the evidence bearing upon this question, even so far as regards the nebula we are now discussing. Not many years ago a new star was observed in the nebula, and the difference between the spectra of the new star and of the nebula itself was merely the addition of the lines of hydrogen! On the meteoritic hypothesis this is easily explained by an increased number of collisions lasting for a time: on the gaseous hypothesis an explanation is not so easy.

If it be granted that we are really dealing with streams of meteorites, all the new phenomena revealed to us by Dr. Roberts' photographs receive a simple and sufficient explanation, especially the apparent condensations here and there which are not condensations of matter necessarily, but *loci* of greater disturbances caused by crossing streams.

The next considerable revelation was obtained from the photograph taken in 1889, of the spiral nebula—long recognised as such—in Canes Venatici, certainly one of the most wonderful spiral nebulæ in the heavens. It is all the more striking because this is a nebula which we look down upon; we see it in plan; we are, so to speak, at the pole of the system, so that it is not foreshortened.

There is no question about the wonderful spirals being connected with the central condensation and stretching towards it. I call attention to the points of condensation along one of the spiral branches, and where we get the possible intrusion of two spirals, one on the other, we see a confused mass of light.

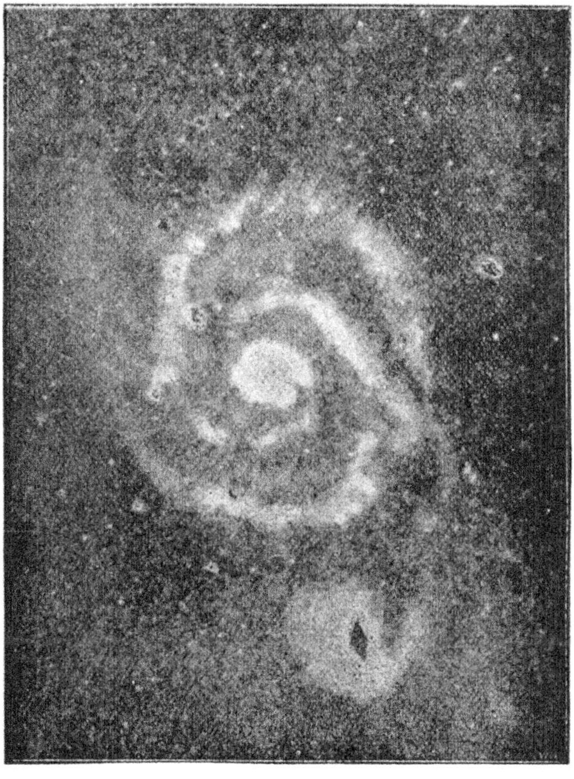

FIG. 41.—The spiral nebula in Canes Venatici, from a photograph by Dr. Roberts, 1889.

Now, if we imagine ourselves dealing there with a mass of pure gas, whether it is hydrogen or nitrogen or ammonia—that is, a combination of both—or any other, it would be extremely difficult to see why there should be any change of temperature in different parts of that mass; but the moment we assume that we are dealing with cool materials—meteoritic dust—we see that such a picture as this is important, for the reason not that it shows us what is there, but because it shows us what is going on there, as already pointed out in relation to the nebula of Andromeda. The bright spots do not represent the presence of matter merely, and the dark ones its absence; but the brighter portions represent the intersection of stream lines where collision is possible—the intervals those regions where collisions are less likely. We can gather from the very configuration of this system that if all the dust, or meteorites, or conglomerations of particles, whatever they may be, are going the same way, there will be a condition in which we shall get a minimum of collisions and, therefore, a minimum of temperature. If the movements are quite orderly and in the same direction, we must not expect to get any very great disturbance, and, therefore—if these disturbances produce high temperatures—we shall not expect to get indications of any particularly high temperatures.

The important point is that here we get apparent stars arranged along the spirals, and we are here in presence of a *vera causa* of the relation of nebulæ to so-called "stars."

Dr. Roberts writes as follows:—

"The photograph shows both nuclei of the nebula to be stellar, surrounded by dense nebulosity, and the convolutions of the spirals in this as in other spiral nebulæ are broken up into star-like condensations with nebulosity around them. Those stars which do not conform to the trends of the spirals, have nebulous trails attached to them."[1]

Strikingly similar to the above is the photograph of M 74 Piscium taken by Dr. Roberts in December, 1893; and here, again, it is also a question of apparent stars.

"The photograph shows the nebula to be a very perfect spiral, with a central stellar nucleus and a 15 mag. star close to it on the south side.

---

[1] Roberts' photographs, p. 85.

The convolutions of the spiral are studded with many stars and star-like condensations, and on the north preceding side there is a partial inversion of one of the convolutions, which conveys the idea of some irregular disturbing cause having interfered with the regular formation of a part of that convolution."[1]

In Messier 101 Ursæ Majoris, which was photographed in May, 1892, we have another case in which the convolutions are broken up into star-like condensations.[2]

I wish to point out that from the centre of the condensation the luminosity gradually gets less and less, until at last we have no luminosity greater than that of the surrounding sky. In the nebula itself we find exquisite spirals, starting apparently from different points, and gradually coming towards the

FIG. 42.—Messier 74 Piscium, 1893.

[1] *Monthly Notices*, vol. liv, p. 438.
[2] Roberts' photographs, p. 89.

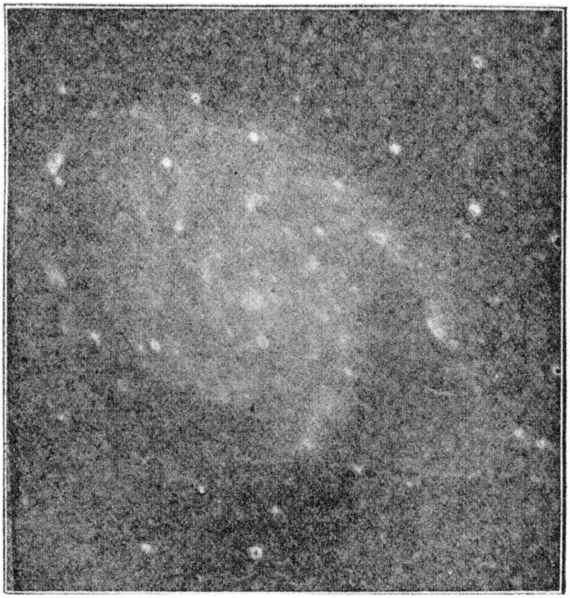

FIG. 43.—Spiral nebula, Messier 101 Ursæ Majoris.

centre, and if we look along these spirals we see that the star-like masses, *which may not be stars*, are in many cases located on the spirals, representing apparently minor condensations, each itself being probably brighter than the other parts because it is more disturbed.

So much, then, for the autobiographical records we now possess of some of the most perfect spiral nebulæ in the heavens.

We see that they all resemble perfect *eddies* in appearance; the question arises, are they perfect eddies in fact? On the meteoritic hypothesis they may well be so, for if moving streams of meteorites encounter resistance to their motion due to disturbances by other masses, the sheets of meteorites are bound to behave like sheets of water; in any case, the *onus probandi* lies with those who hold the contrary view. But in these celestial maëlstroms there are bound to be smaller eddies; and, if Swift had had the opportunity of studying Dr. Roberts' photographs, a more grandiose image might have replaced that in his well known lines—

> "So naturalists observe, a flea
> Has smaller fleas that on him prey,
> And these have smaller still to bite 'em,
> And so proceed *ad infinitum.*"

The question is well worth asking, not only to enable us to explain the photographic and spectroscopic phenomena, but also because we seem to be in presence of forces which must ultimately result in a true star *with rotation,* a concomitant of star life which is not easy to explain, and which Lord Kelvin has shown would certainly *not* be produced by collisions of two finished cosmical bodies.[1]

There is another spiral nebula, however, which may carry us a little further along the same line.

In M 33 Trianguli we have something apparently different from those that have preceded; so much so, that Dr. Roberts, who till quite recently has remained silent with regard to the physical origin of the more regular spirals, has suggested that we may here be in presence of meteoritic collisions.[2] He writes:—

"It will be observed that there are two large, very prominent spiral arms, with their respective external curvatures facing north and south, and that the curves are approximately symmetrical from their extremities to their point of junction at the centre of revolution, where there is a nebulous star of about tenth magnitude with dense nebulosity surrounding it, and elongated in *north* and *south* directions. Involved in this nebulosity are three bright stars and several faint nebulous stars; the two arms also are crowded with well-defined stars and faint nebulous stars with nebulosity between them; and it is to the combined effect of these that the defined forms of the arms are due. Besides these two arms there are subsidiary arms, less well defined, and likewise trending

---

[1] *Proc. Roy. Inst.*, vol. xii, p. 15.
[2] *Monthly Notices*, vol. lvi, p. 70.

FIG. 44.—M 33 Trianguli.

towards the centre of revolution, and are constituted of interrupted streams of faint stars and nebulosity intermingled together; many of the stars are nebulous, and many are well defined but small. The interspaces between the convolutions are more or less filled with faint nebulosity, having curves, rifts, fields, and lanes, without apparent nebulosity in them. They are like the interspaces in clouds of smoke, and cannot be classified.

"There are outliers of nebulosity with many small well-defined and nebulous stars involved in them, and there are also isolated nebulous stars on the extreme boundaries of the nebula; but the evidence is strong that they are all related to the nebula.

"It is by the study of the photographs, and not by descriptive matter, that we can form a true conception of the character of this nebula; from which we shall be justified, even now, in drawing some inferences as to its formation and further developments. To this end I may be permitted to suggest the following.

"We know, with a reasonable amount of certainty, that both nebulous and meteoric matters exist in space; and we also have some evidence that bodies in space have come into collision.

"From these premises we may infer that this nebula is the result of a

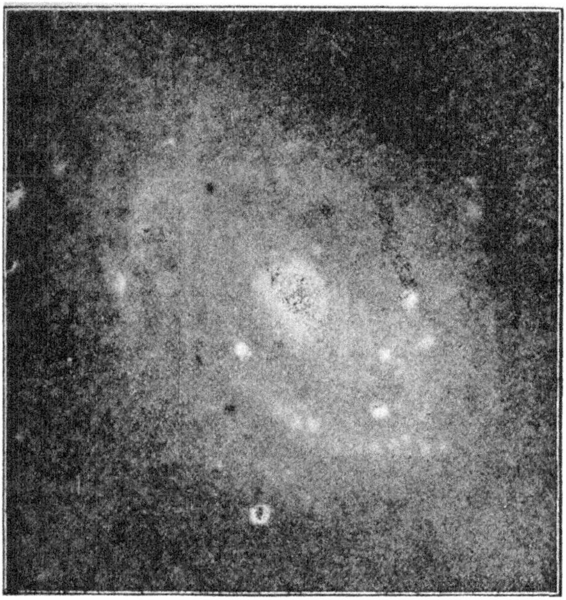

FIG. 45.—H 84 Comæ.

collision of some kind; and we can imagine collisions of at least three kinds possible, namely, (1) between two stars, (2) between two nebulæ, (3) between two swarms of meteorites.

"In the case of this nebula, which (if any) of the three possibilities mentioned seems to us the one most probable to have happened? Much might be said in favour of each of these suggestions, but I shall not at present enter into details, though I think we could readily imagine that the collision of two swarms of meteorites, moving in opposite directions one from the *south following* and the other from the *north preceding*, would account for the spiral appearance, the rotatory motion, and the smashed and scattered state in which the nebula is shown to us upon the photographs."[1]

It is worth while to point out, in connection with the argument in favour of the meteoritic nature of spiral nebulæ, that there are other nebulæ representing streams in space, to which it seems almost impossible to attribute a purely gaseous origin.

[1] *Monthly Notices*, vol. lvi, p. 70.

This branch of work is so young, that there has not yet been time to bring a crucial test to bear on these "stellar condensations" to which reference has been made. *If they could be shown to be short-period variables, then their true stellar nature would be at once negatived.* In fact where spectrum analysis is impossible of application in consequence of the faintness of the stars, variability may help us to point to which bodies the faint stars are most akin.

We have already in two instances obtained important evidence on this point. In 1889, Dr. Roberts was good enough

FIG. 46.—Nebula near 52 (*k*) Cygni.

to allow me to enlarge a photograph of the nebula of Orion, on which there had been a double exposure. I pointed out to Dr. Roberts that variability in some of the stars was suggested; although the exposure was a double one, some of the images were single, and there were *inversions* in the intensities of the double images. Dr. Roberts made a minute examination, with the following results:—[1]

"On examination of the dual stellar images on the photograph the eye immediately detects that ten of them have undergone considerable change in brightness or magnitude during the interval of five days which elapsed between the two exposures. In three of the ten stars, the brightness has increased to the extent of from one-fourth to one-third the measured diameter of the stellar photo-image, and one star appears on the second exposure where none is shown on the first exposure. Six of the ten stars have diminished in brightness during the interval to the extent of from one-fourth to four-tenths, the measured diameter of the photo-image." . . . .

"I have with due care examined the film of the negative under the microscope in order to see if any defect or evidence of defective sensibility on parts of the film could be traced so as to account for the variability in the brightness of the stellar images, but I could not find any such evidence, and I would of course have repeated the photographic experiment if the state of the sky at any time during the past twelve months had permitted. Those who possess the necessary telescopic power may study by eye observations the variability in these stars, and it is one of the functions of the photographic method to point out where eye observations can with advantage be applied in search for special knowledge, and these ten stars are now indicated for that purpose."

The next cases are those afforded by the variability of stars in some globular star-clusters recently photographed at Arequipa with the 13-inch Boyden telescope. An extraordinary number of variable stars was discovered. The *Harvard Circular*, No. 2,[2] states:—

"At least eighty-seven of the stars in the cluster M 3 (N.G.C. 5272), in Canes Venatici, have been found to be variable, and in some cases the change of light amounts to two magnitudes or more. In the cluster M 5 (N.G.C. 5904), forty-six variables were found, out of 750 stars examined, so that they form about 6 per cent. of the whole; of the 16 stars, contained in a circle 110″ in diameter, six are variable. Smaller numbers of variables have been found in other clusters, but in other cases not a single variable has been detected out of the hundreds of stars which have been photographed; the conditions of the search, however, not taking

[1] *Monthly Notices*, vol. l, p. 316.
[2] *Nature*, November 28, 1895.

account of long period changes. In general, no variables have been found within about one minute from the centres of the clusters, on account of the closeness of the stars, and none are more than ten minutes distant from the centres. Some of the newly discovered variables have short periods, in some cases of only a few hours. Thus, five photographs of N.G.C. 5904, taken at intervals of an hour on July 1, 1895, give for the magnitude of a star about three minutes of arc preceding the centre of the cluster, 14·3, 13·5, 13·8, 13·9, and 14·3; four plates, taken at similar intervals on August 9, gave the magnitudes 14·2, 14·6, 14·8, and 15·0."

A special investigation has since been made of the variables forming part of the cluster M 5 Serpentis, N.G.C. 5904 (*Ast. Nach.*, 3354). Forty-five photographs of this cluster have been measured by Miss Leland, and the measures include the greater portion of the forty-six variables previously discovered. The periods of these variables are in general very short, not exceeding a few hours. One of these, designated No. 18, which follows the centre of the cluster about 6′, and is south 5′, has a probable period of 11 hours 7 minutes 52 seconds, or 0·4638 days. The co-ordinates of the light curve of this variable are as follows:—

Days.		Mag.	Days.		Mag.
0·00	....	13·50	0·25	...	14·73
0·05	...	13·87	0·30	....	13·73
0·10	....	14·35	0·35	....	14·72
0·15	....	14·70	0·40	....	14·65
0·20	....	14·72	0·45	....	13·56

It thus appears that the star remains about minimum brightness during half the period, while the maximum luminosity is of relatively short duration; the decrease in light is rapid, but the rate of increase is still more rapid, as it should be. The succession of changes does not seem to correspond with those of any previously known class of variable stars.[1]

Now, since the presence of real nebulous material in some star clusters is accepted by many authorities, there seems ground

[1] *Nature*, June 4, 1896.

for ascribing the phenomena in the nebula of Orion, and in the star clusters, to the same cause, and in attributing them to mere star-like appearances due to collisions. A variability of the kind described, extending over a few hours or a few days, is to me unthinkable in a "star," properly so-called, that is, a body like our sun, and I have no hesitation in expressing my firm conviction that such variability can only be simply and sufficiently explained by the cause assigned for it by the meteoritic hypothesis—a clashing together of streams of meteorites.

If the evidence that the apparent stars are really denser and more disturbed meteoritic swarms is accepted, the view that the nebulæ are gaseous must fall to the ground, because the denser material of the "stars" must be the same as that which was least dense, that is, sparse in the first instance.

I am glad, finally, to be able to state that Dr. Roberts, to whose continuous activity and marvellous skill the world of science is so much indebted, in a paper read at the meeting of the British Association at Liverpool last year, stated his opinion that the origin of the various star-like condensations in the spiral nebulæ is "more probably" meteoritic than gaseous in its origin. The line of argument which has led him to this conclusion will be gathered from the following brief analysis of his communication:—

He draws attention to the remarkable groups, curves, and lines of stars that are clearly shown upon a photograph of the sky in the constellation Auriga, which was taken with an exposure of the plate during ninety minutes. Some of them are constituted of bright stars of nearly equal magnitude; some are of faint stars, also of nearly equal magnitude; some are of both bright and faint stars, and there is much regularity in the spacing distance between the stars in the several groups. These appearances are persistently found upon all photographs taken with long exposures of the plates in any part of the sky where the stars are numerous, such as Cassiopeia and Argo.

"What explanation," he asks, "can be offered to account for the grouping of the stars other than the assertion that they were from the beginning so placed?" He then brings forward the evidence furnished by the spiral nebulæ similar to that I have given above, and which I brought together more than two years ago.

He then goes on:—

"I would submit that the evidence, part of which has now been laid before us, is reasonably conclusive that some, if not many, of the stars which we see in curves and in groups strewn over the sky have been formed in the manner which I have pointed out. There are, besides this, other methods of stellar evolution pointed out on other photographs, such as condensations into stars of nebulæ which have not, at present, symmetrical structures and of globular and annular nebulæ. . . .

"The question will naturally present itself to us: "If it be true that stars are evolved from spiral and other forms of nebulosity, whence came the nebulous matter! We can answer with confidence that it exists very largely and over extensive areas in many parts of the sky; and that it exists there in the form of gas, or, more probably, as Professor Norman Lockyer urges in his *Meteoritic Hypothesis*, of meteors or meteoritic dust."

Dr. Roberts' reference to my work is very encouraging, since there are few workers in science whose researches have so close a bearing on the views I have put forward. For my own part I feel that the totality of the observations above recorded is all highly suggestive of meteoritic action, and, moreover, explains the close relation of "stars" and nebulæ by showing that the former are not formed bodies like our sun; and I can only, in conclusion, express my belief that, before very long, as striking evidence of variability will be found among the stars in the spiral nebulæ as the Harvard observers have obtained from the globular star clusters.

Replies to Objections.

I next turn to some objections which have been raised by Dr. Huggins, who apparently does not accept the view now completely established by Pickering, Keeler, and Campbell, though I confess it passes my comprehension how he can fail to accept it when he has already accepted the major premise of "early evolutionary forms."

The objections have been twofold. First he denied the existence of nebula in connection with some of the "stars" in question, and next he denied the validity of the evidence suggesting the presence of carbon bands in their spectra.

I take the first objection first.

Since I hold these "stars" to be merely the condensing centres of nebulæ, it early occurred to me that possibly by those new methods of inquiry to which I have already referred we might be enabled to demonstrate the existence of the nebulæ, although we can never hope to see them by the unaided human eye. The idea was suggested that long exposed photographs might give us stars surrounded by nebulæ. So I wrote to Dr. Roberts, who always kindly places himself at the disposal of any student, and asked him if he would be so good as to photograph that region of the heavens in which most of the bright-line stars have been observed. He at once acceded to my request, and took photographs, as desired, with his instrument, giving an exposure of three and a quarter hours. The result a little disappointed me, because he reported that there was no indication whatever of any nebulosity surrounding these stars.

I was content to wait for "more light," but Dr. Huggins, who made a similar appeal to Dr. Roberts, with the same result, felt himself justified in objecting to the view which associated those stars with nebulous surroundings. But that is not the whole story. Some time afterwards, at the request of Mr. Espin, Dr. Max Wolf, who has an instrument which is even more competent to pick up faint nebulæ than the wonderful telescope employed by Dr. Roberts, also took photographs of this same region; and I need not say that, being anxious to carry the inquiry as far as he could, he made the exposure what we should consider almost impossibly long—so long, in fact, that one whole night was not sufficient. His first photograph of this region was exposed for thirteen hours on three nights; the next one was exposed for eleven hours.

The results of Max Wolf's inquiries indicated that Dr. Roberts' negative conclusion was probably due to the too great focal length of the telescope employed. I give Max Wolf's results from an American journal.[1]

"Dr. Wolf has photographed the same region with his short focus portrait lens, and although the exposure was too short to bring out any definite form, there seems to be no doubt that the Wolf-Rayet stars, and also the bright-line star P Cygni, are directly connected with nebulous matter. Considering the peculiar type of spectrum which these stars possess, and the fact that they tend to group themselves in the plane of the Milky Way, this discovery of nebulous appendages must be regarded as of considerable importance."

Here, then, we have a perfectly independent line of research indicating that the objection of Dr. Huggins is not valid, and that I was perfectly justified in stating that these bright-line stars were associated with nebulæ. The statement first made on theoretical grounds is now backed up by exquisite photographs, which indicate that most certainly there is a complete association of nebulous matter with these stars.

I must here point out the enormous advantage students of science now have in possessing such magnificent photographs as these. Not only is the wealth of science rendered obvious, but the wealth of nature; in them we have what modern science makes of a little patch of the sky on which the naked eye sees nothing at all.

In a photograph of the region surrounding the brightest star in the constellation Cygnus, we have here and there indications of nebulous matter as well as of stars. That is rendered evident by the fact that in certain other regions we get a perfectly

---

[1] *Ast. and Ast. Phys.*, March, 1892, p. 236.

dark back-ground, whilst in others the back-ground itself is luminous. In the region in which these bright-line stars have been recorded for several years it is almost impossible to point out a large area in which there is not a most obvious indication of luminous nebulosity. Patches here and there seem to indicate that the great differentiation between this part of the sky and others, lies not in the wealth of stars, but in the wealth of the luminosity in which they are situated.

I next take the question of carbon. The spectrum of carbon is one which is subject to very great changes when examined under different experimental conditions, and an acquaintance with these variations is essential to an adequate discussion of the spectra of the heavenly bodies.

Everyone knows the considerable development of the spectrum of carbon in most cometary phenomena; and there is a band which Dr. Vogel some years ago attributed to carbon, although it does not coincide with the most familiar carbon spectrum, that of the Bunsen burner. Dr. Vogel gave his reasons for this allocation, and illustrated them by a diagram[1]

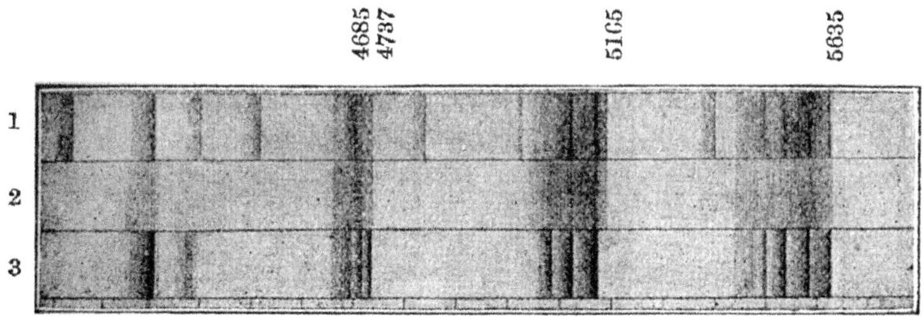

FIG. 47.—(1) Spectrum of mixture of hydrocarbons and oxy-carbon obtained from meteorites, with small coil and jar. (2) Spectrum of Comet III, 1881. (3) Spectrum of gases from meteorites, with large coil and jar (Vogel).

in which it is shown that in the spectrum of the comet, the blue band has its maximum about 470, and fades away nearly equally in both directions. If one's knowledge of the carbon

[1] *Potsdam Observations*, 1881, vol. ii, p. 173.

spectrum were limited to that of the Bunsen burner, indicated at the bottom of the diagram, the comet band could not be ascribed to carbon.

But another spectrum of carbon, obtained by Dr. Vogel (given at the top of the diagram), shows the blue band of exactly the same form, and in the same position as that in the comet. Hence, Dr. Vogel argued that the blue band in the comets, though not coinciding with the carbon group at wave-length 4737, seen in a Bunsen flame, was still due to carbon.

When I was discussing the spectrum of the bright-line stars in my general survey, a band in very nearly, if not absolutely the position of the cometary band, was found recorded. Most unfortunately I had completely forgotten Dr. Vogel's paper of 1881, and I set to work to study its origin for myself.

In the course of the previous thirteen years I had taken some hundreds of photographs of the spectrum of carbon compounds under a great variety of conditions; and I was driven to carbon because one of the most conspicuous features of the spectrum of many of the bright-line stars is a broad blue band, a part of which falls within the limits of the group of carbon flutings at 4737, which is seen in the Bunsen, although its brightest part in the stellar spectra is about wave-length 468; that is, some distance further towards the blue than the brightest part of the Bunsen group. Forgetting Vogel's prior labours, I looked through my photographs, and found what he had found, that, under certain conditions, the maximum of the band is shifted, under some conditions of pressure and temperature, from 4737 to about 4685. One of my photographs, taken in December, 1886, is reproduced in Fig. 48.

The spectrum at the top of Fig. 48 is the spectrum of alcohol vapour at a relatively high pressure; and among the most notable differences from the lower pressure spectrum at the bottom of the diagram, is the enhancement of the more refrangible part of the group of flutings commencing at 4737. This

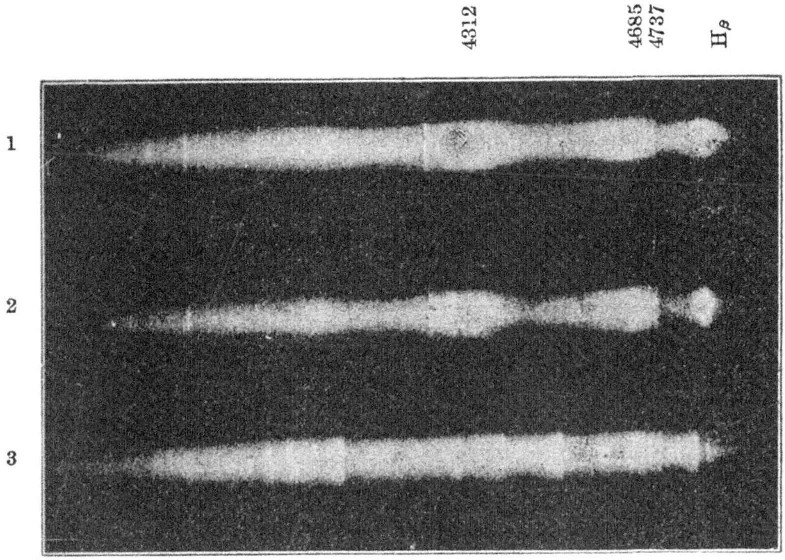

FIG. 48.—Spectrum of alcohol vapour. (1) Highest pressure. (2) Intermediate pressure. (3) Lowest pressure.

is still more marked in the spectrum at intermediate pressures, as shown in the middle spectrum in Fig. 48. In short, the brightest part of this group now agrees, very nearly, with the blue band of some of the bright-line stars. It is very difficult to estimate the middle of such a broad, diffuse band as that in question; but the wave-length of the brightest part is approximately 4685.

As Dr. Vogel had previously done for comets, I was particularly careful to point out that the carbon band in the bright-line stars was not seen under the same conditions as that in the Bunsen flame.[1]

[1] In 1888 I wrote: "This band is evidently the bright band of carbon, commencing at 474, with a maximum about 468, as observed and photographed at Kensington" (*Proc. Roy. Soc.*, vol. xliv, p. 35). Later in the same year I added: "It is necessary to state that the maximum luminosity of the blue band under some conditions, is about 468 . . . . The conditions under which this band has its maximum luminosity at 468 in Geissler tubes seem to be those of maximum conductivity" (*Proc. Roy. Soc.*, vol. xlv, p. 167).

I next append some observations of the band in both comets and bright-line stars:—

Comet.	Maximum of band.	Observer.	Star.	Brightest part.	Observer.
Winnecke. 1868	469	Huggins	No. 3821	469	Huggins.
Comet IV. 1873	469	Vogel			
Comet III. 1881	468·5	Vogel	No. 4001	468	Huggins.
Comet III. 1881	468	Copeland			
Comet IV. 1881	470	Copeland			

Notwithstanding the difficulty of determining very exactly the brightest part of a diffuse band, it will be seen that the bands in two of the stars are exactly coincident with bands which have been measured by trustworthy observers in three comets. In the fifth comet, named above, the variation in the wave-length of the band is not greater than that between two individual measures in stars.

The fact that, so far as I know, this explanation, the whole credit of which is due to Dr. Vogel, has never been called in question in relation to comets, would indicate that it may be equally unobjectionable in the case of the bright-line stars.

But there was more evidence behind with regard to the other carbon flutings.

The spectrum of carbon does not consist of the blue band alone, so that some account must be rendered of the other parts of the carbon spectrum, more especially of two groups of flutings in the green. We should not expect the green flutings to be so easily visible as the blue in stars, for the reason that they fall in the brightest part of the continuous spectrum; while in comets where there is little continuous spectrum, they are the most conspicuous bands.

In the star BD + 36° 3956, for example, Vogel's observations gave indications of the two green bands which are seen in the spectra of carbon compounds, and are the chief characteristics of the spectra of comets at mean distances from the sun.

It was in 1890 that Dr. and Mrs. Huggins formulated objections, based on some new measures, to my view as to the probable carbon origin of the blue band in the bright-line stars. Thus, in two cases they found, as Vogel had found before them, that the brightest band was still more refrangible than the brightest part of the modified carbon band; but in each of these cases they found, also, a band about the position of 4685.

As a result of their work, they made the following statement:—[1]

"Our observations appear to us, however, to be conclusive on the main object of our inquiry, namely, that the bright blue band in the three Wolf-Rayet stars in Cygnus, and in DM + 37° 3821, is not coincident with THE BLUE BAND OF THE BUNSEN FLAME." [The capitals are mine.]

It will have been seen how carefully Vogel and myself had pointed out that it was not a question of the Bunsen flame!

Dr. and Mrs. Huggins do not admit that the observed variations of the band in the carbon spectrum are sufficient to explain the position and appearance of the band at 4685 in the stars, basing their objections on experimental evidence afforded by Hasselberg.

Vogel's researches, as well as my own, on the carbon spectrum, however, indicate a much greater concentration of luminosity of the band, about 4685, than appears to have been observed by Hasselberg. But this is not to be wondered at, since every change in the experimental conditions may have an effect on the spectrum.

I am not aware of any other objection to my view than the above, and it will be remarked that Dr. Huggins is silent altogether in regard to the existence of the band in comets.

Very fortunately for science, a great mass of new work on

---

[1] *Proc. Roy. Soc.*, vol. xlix, p. 46.

this part of the subject has been brought together since *The Meteoritic Hypothesis* was published, chief among the workers being Professor Campbell.

Let us turn to this new work, therefore, and see in what direction it tends.

We may take the case of one of the brightest stars of this class in Argo, the spectrum of a star which my friend Respighi and myself were the first to see on a very hot night in Madras in 1871, a beautiful spectrum with many bright lines. Professor Campbell in 1893 included the study of this bright line star in his work at the Lick Observatory. What is his result with regard to the band at 4685? He finds a band at 4688.[1] In my first discussion I took the position of the brightest band as 4682, depending upon measurements made by Ellery at Melbourne in 1879. In a more general examination of all the Wolf-Rayet stars in 1894,[2] he finds a band at 4688 to be the most constant feature, and in some stars it appears almost alone! This in itself would be almost sufficient to prove carbon.

Professor Campbell does not discuss the origins of the lines and bands which he has measured, but it will be seen that for such a diffuse band as that in question his wave-length 4688 does not differ materially from that already given for the modified carbon band.

There is now, therefore, in the light of the newest and best work, no question about the fact that in the bright line stars there is a band at 468, the wave-length of the modified carbon band; and this was my original contention.

The new measures obtained by Dr. and Mrs. Huggins, therefore, do not affect my views as to the origin of the blue band of the bright line stars recorded at places varying between λ 468 and λ 469.

With regard to this band, then, the first conclusion is now

---

[1] *Astronomy and Astrophysics*, 1893, p. 555.

[2] *Ibid.*, 1894, p. 448.

firmer than ever, strengthened as it is by Campbell's new work.

Dr. and Mrs. Huggins object to another point.

It has already been remarked that Vogel's observations suggested the presence of the green flutings in one of the stars.

On this point Dr. and Mrs. Huggins remark that, when observing the bright line stars with small dispersion, "it might easily be supposed that the spectrum is brighter at the position of the green carbon band." With high dispersion, however, they can see "no sensible brightening" in that part of the spectrum. In the case of another line distinctly seen with small dispersion, they remark that with high dispersion it was so indistinct that they could not determine whether it was D or D^3;[1] so that their observations do not demonstrate the absence of the green fluting of carbon.

Observations made at Kensington strengthen the idea that the green flutings are present in the spectra of bright line stars. When high dispersion is employed, flutings are weakened in much greater proportion than lines; so that comparatively small dispersion must alone be employed in observations of this kind.

Campbell shows that, while the average position of the brightest blue band in one group of stars is 4688, in another group it is 4652. These two bands are frequently associated in the same spectrum, but occasionally each occurs by itself.

As to those stars in which a band appears about λ 465, it is quite possible that we may still have to deal with carbon. At present I am not aware of any experimental evidence; but the possibility of a band at this wave-length under a certain still untried condition is suggested by the fact that a band about this position was observed in Brorsen's comet in 1868 and 1879.

---

[1] *Proc. Roy. Soc.*, vol. xlix, p. 44.

But the existence of a band at 465 surely does not negative the existence of a band at 4685!

The present position of the question of carbon, then, is this. The new work of Campbell justifies us in associating, not only in comets, as first suggested by Vogel, but in bright line stars, as suggested by myself, the blue bands at 468 with carbon; and a study of the spectra of comets suggests, but does not demonstrate, that the other band at 465 has the same origin. The feeble appearance of the green bands is no doubt due to their superposition upon the brightest part of the continuous spectrum.

Hence the idea of the chemical and physical kinship of comets, nebulæ, and bright line stars is strengthened.

CONCLUSION.

Finally then I would submit, in bringing this part of the inquiry to a conclusion, that all the new observations have strengthened the views to which I was led by a study of the old ones; while, on the other hand, the only objections I am aware of brought against these views have been rendered invalid by the progress of the inquiry made by other workers.

Max Wolf has shown, beyond all question, that some of the bright line stars are associated with nebulæ; Campbell has shown that many bright line stars have a band in their spectra agreeing in wave-length with a band observed in cometary spectra; while, finally, both Vogel and myself have found that this band is in all probability due to carbon.

CHAPTER X.—DOUBLE SWARMS.

The New View of Variables.

The next part of the inquiry lands us among phenomena which so far have been considered to be exceptional. I refer to the so-called variable stars.

If there be any truth in the view we are considering, it will be abundantly clear to every one that the light of stars as they pass from the nebulous to the more luminous stage must change during the progress of that evolution. But that change will not be visible to one generation of men, probably not to a thousand generations of men. It is a change which will require millions, and possibly billions, of years for its accomplishment; and, therefore, we must not associate the word "variable" with any change which depends wholly upon the evolution of these various stellar conditions. But, in addition to that, we can see almost in hours, certainly in days, frequently in months, sometimes in years, changes in the light of certain stars; and it is these short period changes which mark out and define for us the phenomena of variable stars.

I am not without hope that in future years the explanation afforded by the meteoritic hypothesis of the light changes thus observed will be acknowledged to be one of the most important parts of it. So long as all stars were considered to be bodies like the sun, the various kinds of variability observed were hard to explain. The grotesqueness of some of the explanations may be gathered from the statement I have formerly given. But the idea that all stars are *not* like the sun, but, on the contrary, may exist in different degrees of condensation of meteoritic swarms, at once opened the door to previously undreamt-of possibilities in the causes of light change.

I wrote in 1881[1]:—

"A possible explanation of most of the new and variable stars is to be found in the meteorite theory; the innumerable components of one group of meteorites colliding with those of another group would be competent to give out light sufficient to make the whole appear as a star. Each meteorite gives only a little light, but the total must be very considerable."

Limiting myself here to regular variability, I may, as a reminder, state how it was treated in the meteoritic hypothesis. The general statement made was as follows[2]:—

"*Regular Variability.*

"All regular variability in the light of cosmical bodies is caused by the revolution of one swarm or body round another (or their common centre of gravity).

"In the case of the revolution of one swarm round another, an elliptic orbit is assumed, and the increase of the light *at maximum* is produced by collisions among the meteorites at periastron.

"In the case of the revolution of a swarm round a condensed body, the increase of the light *at maximum* is produced by the tidal action set up in the secondary swarm.

"In the case of one condensed body revolving round another, the reduction of the light *at minimum* is caused by an eclipse of one body by the other. This can only happen when the plane of revolution of the secondary body passes very nearly through the earth."[3]

The great differences in the phenomena to be expected, then, on the hypothesis must depend upon the degree of condensation of two bodies; *and there must always be at least two bodies.* No short period variability can, I hold, take place in one body alone. Take a star like the sun. It is pretty obvious that any change in the sun, such as we see it now, would require a very con-

[1] *Encyclopædia Britannica*, vol. xxii, p. 651.
[2] *Meteoritic Hypothesis*, p. 475.
[3] *Ibid.*

siderable time for its accomplishment, so as to be obviously visible to us all; but, if we consider two bodies like the sun, we can imagine a condition of things in which one body would come exactly in the line between the earth and the other body, and would so eclipse the further one. There we have at once the possibility of an eclipse due to the passage of one condensed body in front of another, and, therefore, of a variability which depends upon eclipses. So much for two bodies like the sun; but we know that in various parts of celestial space some of the stars have run through their life of light, and exist as dark bodies. Obviously we should get the same eclipse phenomena when dealing with one star like the sun and another dark, condensed body, provided always that the dark body came and eclipsed the light one. That is a very well known and accepted cause of variability, and one of the most obvious cases of this kind we have in the star Algol. There we have two bodies a bright and a dark one. When we come to examine the light curve of a body like this we find that the luminosity of the star remains constant for some considerable time in relation to the period of variability, and then it suddenly decreases. It almost at once—in an hour or two—goes up again, continues then for another period, and suddenly diminishes again.

Spectroscopically we can inquire into the question as to whether there is or is not any physical change connected with this. Obviously, if it is merely an eclipse, there should be no physical change, and, therefore, no change in the spectrum.

This has been done by Professor Pickering, who has obtained photographs of the spectrum of this star when it was most luminous and when it was least luminous, and the spectra of these two conditions are quite similar. The broad lines are alike; in other dark lines also there is no change. Therefore spectroscopically we are justified in saying that the theory that variability is caused in condensed bodies like Algol by eclipses is perfectly justified.

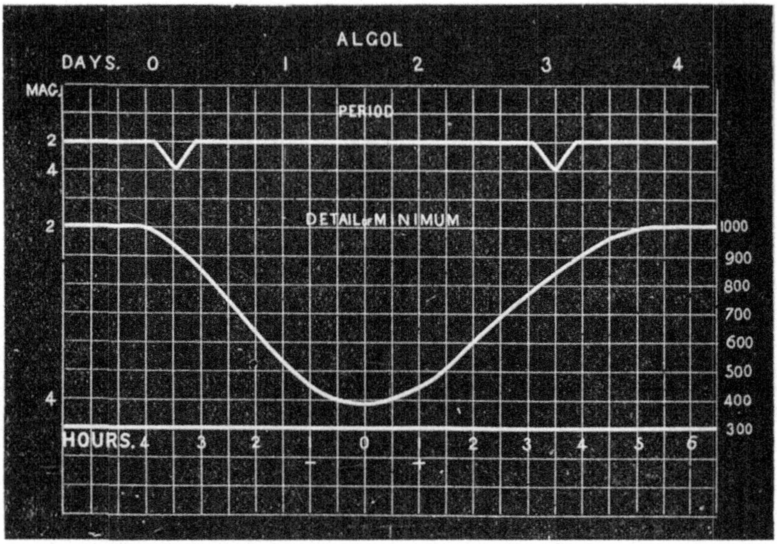

FIG. 49.—Light curve of Algol.

But suppose we consider, not two bodies like the sun or even one sun and another body more condensed and colder than the sun, but two not completely condensed meteoritic swarms, various probabilities never before considered lie open to our inquiry.

We may suppose that round a denser and larger one a smaller one is moving in the orbit represented in Fig. 50. We see that for a considerable part of the orbit the smaller swarm can perform its movement along the orbit without any chance of running up against any of the constituents of the greater swarm; but when the smaller swarm passes through what is called the periastron point—*i.e.*, the region nearest the centre of gravity—which is occupied by the densest portion of the primary swarm, it is impossible that it can get through without a considerable number of collisions between its own constituents and the constituents of the major swarm. What will happen? We get light and heat produced, a variable star is born, which will give the greatest amount of light when the two swarms are closest together.

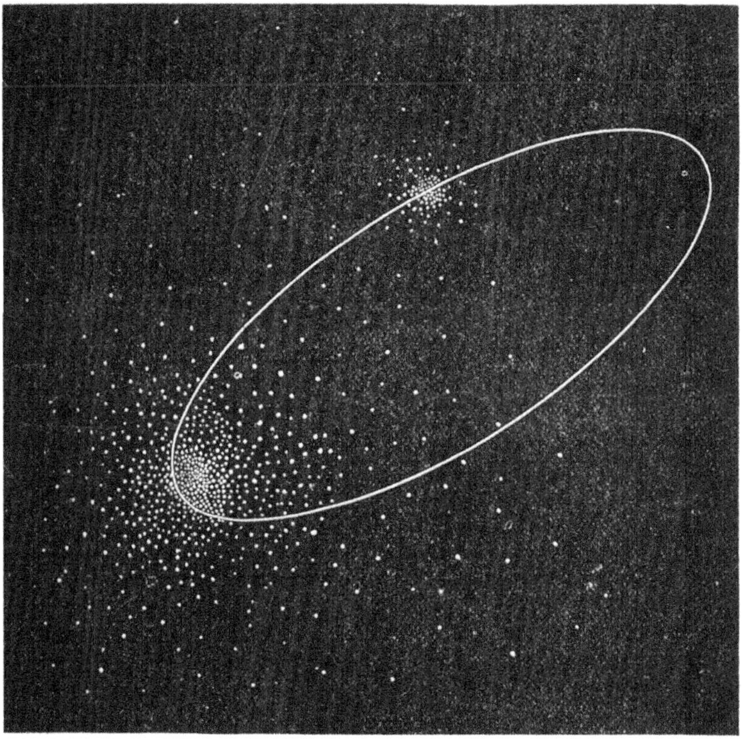

FIG. 50.—Cause of variability in uncondensed swarms.

The minimum, or least, amount of light and heat radiated will occur, not at the time of apastron passage, as would at first be supposed, when the two swarms are at their greatest distance from one another, but some time after the maximum depending on the magnitude of the disturbances set up at periastron passage.

We can imagine also that, instead of dealing with a highly elliptic orbit, such as imagined in Fig. 50, we may have one in which the main mass is very much nearer the centre of the orbit of the smallest swarm, that orbit being much more circular than in the former case. There we get a chance of a number of collisions, extending over a greater part of the orbit of the minor swarm; but there will not be anything like so great a difference between the number of collisions at the two ends of the major axis of the orbit as there would have been in the first case supposed.

In that way, therefore, we can explain the variability of these uncondensed swarms, and not only the variability, but a very considerable difference in the time occupied by the changes and in the intensity of the greatest light produced. So much was that to be anticipated, that I predicted in 1888 that when we got any indication of stars the spectra of which showed that they were really sparse swarms, such as that depicted on the diagram, at the maximum of their luminosity we should get bright lines, and in all probability bright lines of hydrogen, visible in their spectra; in short, that *o* Ceti, otherwise Mira, in the spectrum of which Pickering had photographed bright lines in 1886, would be a typical, and not a special, case.

In *The Meteoritic Hypothesis,* I brought together a considerable number of observations in support of my views, and they have been repeated since, but it is not necessary to go into details. Indeed, it is recognised thoroughly now that one of the chief characteristics of a large class of variables represented by Mira is the appearance at the time of maximum, at the top of the light curve, of the bright lines of hydrogen.

Mrs. Fleming, who is in charge of the work on variables at the Harvard College Observatory, writes:—

"If an object is detected on any of the photographs showing a spectrum of the third type, having also the hydrogen lines bright, it is at once suspected of variability, since only variables of long period are known to possess this peculiarity."[1]

I should add in explanation that the third type of stars

---

[1] *Astronomy and Astrophysics*, 1893, p. 685.

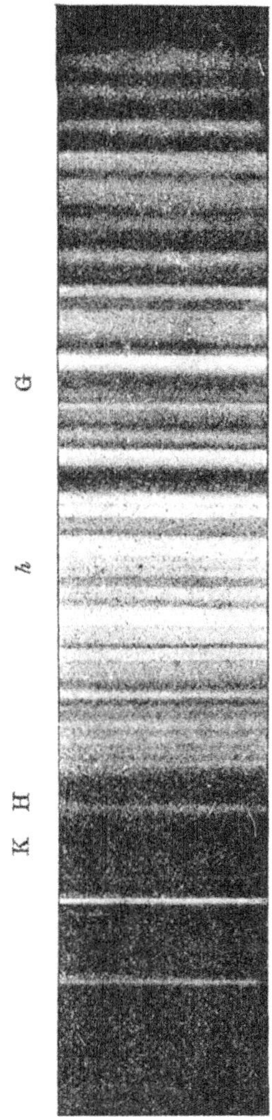

FIG. 51.—Spectrum of *o* Ceti (Pickering), showing the bright lines at maximum.

referred to are stars which I hold to be meteoritic swarms simply, *i.e.*, not condensed bodies like the sun.

Professor E. C. Pickering, in summing up the Harvard work, also states[1] :—

[1] *Astronomy and Astrophysics*, p. 721.

"This property has led to the discovery of more than twenty objects of this class, and no exception has been found of a star having this spectrum whose light does not really vary. Dealing with the variables of long period which have been discovered visually, the hydrogen lines have been photographed as bright in forty-one, the greater portion of the others being too faint or too red to be studied with our present means."

At maximum forty old variables of this class show bright lines, and twenty new variables have been detected by the appearance of bright lines; *i.e.*, bright lines being seen in them suggested that they were variable, and a further inquiry into the old records showed that undoubtedly their light had varied.

The Variable β Lyræ.

I now pass on to consider such a variable at β Lyræ, concerning which star much work has been done since my book was published.

The light curve of this star is very curious. At its lowest brightness it is a 4½ magnitude star, and at its greatest brightness it shines as a star of 3½ magnitude, the changes going through one magnitude. These changes are run through in a period of thirteen days. But the interesting point is, that half-way between the succeeding principal minima, which occur every thirteen days, there is a well-marked secondary minimum at which the light only descends to 3¾ magnitude. These facts will be gathered from the accompanying light curve.

Now what was the old explanation of these curious changes? It was that the star represented a surface of revolution, the ratio of the axes being 5 to 3, *i.e.*, elliptic beyond any experience of ours with regard to any other bodies. There is a dark portion at one end of the axis symmetrically situated. This body then has to turn and twist with its axes and the black spot, and so on, so that at the end of the chapter such a light curve as that of β Lyræ is produced. Now on the new view this variability is produced by movement and by two bodies.

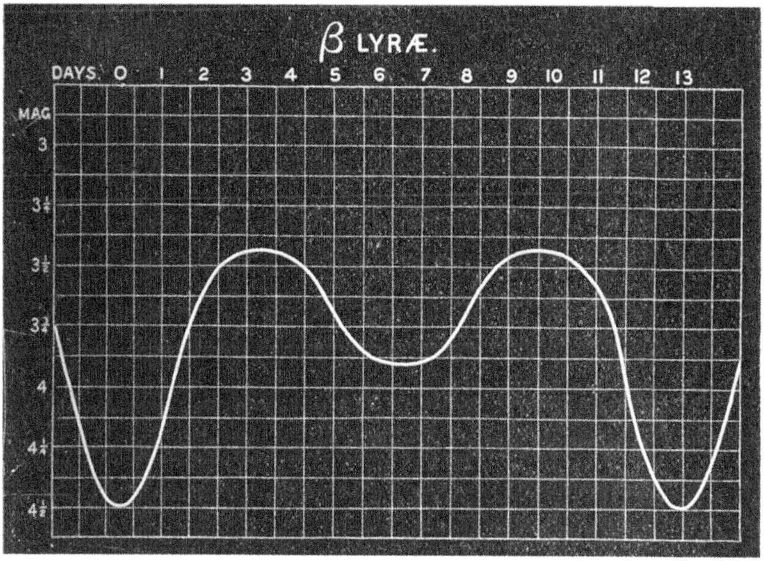

FIG. 52.—Light curve of β Lyræ.

Let us attempt to get at what really happens by examining the spectrum observations.

Pickering was the first to show that β Lyræ was a spectroscopic double, and it was the first spectroscopic double which was discovered in which the companion was not dark. The duplicity was discovered by the shifting of the lines, due to motion of the two bodies in the line of sight. Here, then, was a demonstration of movement and of two bodies.

Next it was found that the spectrum, like that of Mira, contained bright lines, hence the possibility of collisions, another requirement of the new view.

Professor Pickering found that during the first half of the period—that is, between principal and secondary minima—the bright lines were on the less refrangible sides of the corresponding dark ones, while during the second half they were displaced to the more refrangible sides. He further remarked that—

"The actual changes in the spectra, when studied in detail, are much more complicated than has been stated above, and show a variety of intermediate phases and changes in the dark as well as in the bright lines."

At Professor Pickering's request, I took up the work at Kensington in July, 1891.[1] Since it was commenced, accounts of the photographic spectrum of β Lyræ have also been published by Belopolsky,[2] Father Sidgreaves,[3] and Vogel.[4] I proceed to state, step by step, the results of the Kensington observations.

1. *The spectrum is constant at the same interval from principal minimum.*

Apart from the slight differences which seem to be accounted for by differences in the atmospheric conditions and consequently in the quality of the negatives, the spectrum appears to be the same at the same interval from minimum.

2. *The kinds of variation shown on the photographs are as follows:—*

(*a*) Periodical changes in the relative intensities of the lines.
(*b*) Periodical doublings of some of the dark lines.
(*c*) Periodical changes in the positions of the bright lines with respect to the dark ones.

3. *There are two bodies involved giving dark line spectra.*

At, and just before and after the second maximum, some of the dark lines are doubled. This indicates two sources of light giving dark line spectra and moving relatively to each other in the direction of the line of sight. When the relative movement

[1] *Proc. Roy. Soc.*, 1894, vol. lv, p. 279, *et seq.*
[2] *Mem. Soc. Spett. Ital.*, June, 1893.
[3] *Monthly Notices, R.A.S.*, 1894, p. 96.
[4] *Sitzungsberichte*, Berlin, February, 1894.

in the line of sight is zero, none of the lines are doubled. The latter condition occurs about the time of the two minima.

4. *The maximum relative velocity of the two dark line components in the line of sight is about* 156 *miles per second.*

The greatest separation of the dark lines occurs about the time of second maximum, and the relative velocity as determined by measurements of three of the doubles in the photograph of August 24, 1893, is that stated above. The individual measurements are as follows:—

$H\gamma$ = 155 miles per second.
$H\delta$ = 154 "
$\lambda 4026$ = 158 "

5. *One of the dark line components bears a strong resemblance to Rigel, and the other to Bellatrix.*

The spectra of the two components can readily be separated, for the reason that only lines common to both will be doubled. Among these are the lines of hydrogen. Lines special to either component are always single, and they retain the same relative positions with respect to one group of hydrogen lines throughout the period.

In Fig. 53 photographs are given to facilitate an analysis of the compound dark line spectrum. At the bottom of the diagram is a reproduction of a photograph taken near the time of second maximum (August 24, 1893), and the spectra of Rigel and Bellatrix are included in the same plate. The compound character of the dark line spectrum of β Lyræ at this time is shown by the fact that one group of lines corresponds very closely with those which appear in the spectrum of Rigel, and when these are subtracted from the whole spectrum, a spectrum closely resembling that of Bellatrix remains, the latter spectrum being displaced in this photograph to the more refrangible side, as shown by the short lines drawn beneath the spectrum. The two dark line components of β Lyræ may be called R and B respectively.

FIG. 53.—Demonstration of the Components of β Lyræ.

The photographs show that about the time of principal minimum the dark line spectrum of β Lyræ (2) is very similar to that of Bellatrix (1), while about the time of secondary minimum the spectrum of β Lyræ (3) becomes more like that of Rigel (4), the differences at these times being mainly in the intensities of the lines. The photograph of the spectrum about the time of second maximum (5) shows that there are two spectra displaced with respect to each other. The spectrum displaced to the less refrangible side is shown to resemble that of Rigel, while that displaced to the more refrangible side more closely resembles Bellatrix.

It is not intended to suggest that the spectra of the two dark line components are quite identical with those of Rigel and Bellatrix. These are simply the best-known stars which they most closely resemble, and the similarity is pointed out as an indication that we have not to deal with bodies of an unfamiliar type.

The conditions at first maximum are not so simple as those at second maximum, though there is evidence to show that at this point of the light curve the component B is receding with respect to R. The hydrogen lines are broadened, and the two lines near 4472 and 4482 have approached each other, as they should do if one belongs more especially to R, and the other to B.

6. *When the two bodies lie along the line of sight, partial eclipses occur. This happens near the minimum of the light curve.*

The differences in the intensities of the dark lines special to R and B, near the two minima, indicate that near the principal minimum R is partially eclipsed by B, while near secondary minimum B is partially eclipsed by R. If we leave the bright lines out of consideration, it is seen that near principal minimum the spectrum of β Lyræ greatly resembles that of Bellatrix, the component B in this case lying between us and component R. As the eclipse is not total, however, the lines special to R appear with reduced intensities; the lines joining the spectrum of β Lyræ to that of Bellatrix indicate the principal lines of component B. At the secondary minimum, on the other hand, component R lies in front of component B, and the spectrum consequently bears a greater resemblance to that of Rigel. This is shown by the lines joining those of β Lyræ to the spectrum of Rigel in Fig. 53.

The difference is especially noticeable in the case of the lines near λ 4471·8, 4481·8, 4388·7, and in the group of four lines a little less refrangible than H_δ. Near principal minimum 4471·8 is stronger than 4481·8, as in Bellatrix, while about secondary minimum 4481·8 is stronger than 4471·8.

If the eclipses were total, the variations of the spectrum might be expected to be still more striking.

7. *In addition to dark lines, there are several bright ones which change their positions with respect to the dark ones.*

The photographs show conspicuous bright lines about wave-lengths 4862(H_β), 4715·9, 4471·8, 4388·7, 4340·6(H_γ), 4101·7(H_δ), 4025·7, and 3887·7(H_ζ). Other fainter ones also appear in some of the best photographs. Recent research shows them to be due to hydrogen and helium.

The displacements of the bright lines described by Pickering are confirmed in the main by the Kensington photographs. In the several photographs taken between principal and secondary minima, the bright lines lie on the less refrangible sides of the dark ones; at secondary minimum the broad bright lines are almost bisected by dark ones; while from secondary minimum to principal minimum the bright lines are more refrangible than the dark ones. The investigation of the movements of the bright lines must, however, be now carried on in the light of the knowledge gained with regard to the existence of two sets of dark lines.

If we consider the displacements of the bright lines with reference to the dark lines of component R, we find that they are always in the same direction as those of component B with respect to R. Thus, in the first half of the period, the bright lines as well as the dark lines of component B are less refrangible than those of component R, while during the second half they are more refrangible. The bright lines, however, do not keep a constant position with respect to those of component B, although displaced in the same direction.

8. *The bright lines are brightest soon after secondary minimum.*

If the brightness of the lines in reality remains constant, they will appear relatively brightest at the two minima, owing to the reduction of continuous spectrum which is associated with the increased brightness of the star at maximum, and for the same reason they should appear brighter at principal than at secondary minimum. Estimates of the brightness of the lines in relation to the continuous spectrum have been made independently by four of my assistants, and, although estimates of this kind are liable to error, the general agreement is sufficient to indicate that, when all allowance is made for the varying continuous spectrum, there is a maximum of brightness of the bright lines about half a day after secondary minimum. The apparent increase of brightness near principal minimum seems to be due solely to the reduced intensity of the continuous spectrum.

These observations are now four years old; so far as I know, they have never been contested; and I claim for them full confirmation of the three fundamental requirements of the meteoritic hypothesis, for the regular brightening of the lines suggests collisions, and we know now that we have two uncondensed meteor swarms in motion round their common centre of gravity as in *o* Ceti.

Various Kinds of Variability.

It is perfectly clear from the hypothesis that special kinds of variability must accompany each special degree of condensation; *e.g.*, two sparse swarms cannot eclipse each other, nor can they in collision give the same spectrum as when their constituent meteorites are not entangled. Hence, if we find stars with similar spectra associated with similar kinds of variability, we have in this a strengthening of the hypothesis and proof, if any more were needed, that all so-called "stars" are not condensed like our own sun.

Now this is conceded. Professor Pickering writes as follows:—

"The classification of the stars according to their spectra is so far-reaching that it should be applied to each of their other properties. For instance, of the variable stars it appears that all known Algol stars have spectra of the first type (Group IV), while long-period variables in general are of the third type (Group II), and have the hydrogen lines bright when near their maxima."[1]

"Variable stars of short period generally have spectra of the second type" (Groups III or V).

Indeed, we may use the phenomena of variability in support of the general hypothesis, for the different kinds of variability with which we are most familiar must be special to certain kinds of stars. As I have before said, sparse swarms cannot eclipse each other, and it is only sparse swarms which by collision can produce bright lines lasting only a short time.

We should expect, therefore, the variability of stars of which the meteoritic supply has ceased to comprise eclipse phenomena only, and this is so. The more these stars shrink on cooling, the rarer should eclipse phenomena become, and this is so.

Hence we have special kinds of variability for stars increasing and others decreasing their temperature.

So far no mention has been made of variables of the δ Cephei class, but here again it appears that they represent bodies increasing their temperature and less sparse than Mira, while in some respects the conditions are changed.

Replies to Objections.

Two objections, however, have been made to these hypothetical two swarms. It has been urged that the secondary swarm which we saw moving in a closed orbit round the primary one would soon spread out into a line along the

---

[1] *Astronomy and Astrophysics*, 1893, p. 721.

orbit, so that there would always be some parts of it mixed up with the constituents of the parent swarm.[1]

"The conditions in a system of colliding swarms could not be permanent, for with successive collisions the smaller swarm would become spread out so as to form a ring, and hence the variability of the star would cease."

That is a perfectly fair objection, supposing we are dealing with millions and billions of years, but I think that those who have made it do not know the history of astronomy. Let us take, for instance, the history of the well-known November swarm which cuts the earth's orbit, so that in certain Novembers, about thirty-three years apart, we get this swarm of meteorites passing through our atmosphere, getting burnt out in that passage, and giving us one of the most magnificent sights which it is possible for mortals to see: a whole hemisphere of sky filled with shooting stars. Some of us remember such a phenomenon as that in the year 1866; some of us are hoping to see the recurrence of it in 1899, for which we have not long to wait. But the fact that we only get this appearance every thirty-three years shows that, at all events, that swarm of meteorites to which the phenomena are due has not changed during our lifetime; nay, it has not changed during the last thousand years, for man has known of that November swarm for more than a thousand years, and we have only known of the variability of Mira for 300 years. Hence such an objection is entirely out of court, because it lacks the historical touch.

Another objection which has been urged is that there are certain irregularities in the light curves of these bodies which are not consistent with orbital movement; that Mira, for instance, does not always come up to the same amount of brightness at its maximum, and perhaps, for all we know, does not always go down to the same low magnitude when it is at its lowest.

---

[1] Frost and Scheiners, *Astronomical Spectroscopy*, p. 311.

This is Professor Young's objection. He writes:—

"There are many good points about this ingenious theory, but also serious objections to it, as, for instance, the great irregularity of the periods of stars of this class, an irregularity which hardly seems consistent with such an orbital revolution."[1]

That also is perfectly true, and on this account: there is no reason why we should suppose that these phenomena of the waxing and waning light of the body are produced by the movement of one body only. Suppose, for instance, that there is some cosmic eye a billion miles away from our solar system, so beautifully and exquisitely wrought, so delicate in its construction, that it can see an increase in the light of the sun every time a big comet goes round it. Now we know from our own experience of comets that it would be absolutely impossible for that delicately constructed eye to see anything like a constant variability in the light of the sun under these conditions, because sometimes the brightest comets which come to us are absolutely unpredicted; they come at irregular times.

It must next also be pointed out in connection with this objection that there are other obvious causes for considerable variations in the light, both at the maximum and at the minimum. The beautiful spiral nebulæ of which Dr. Roberts has given us such magnificent photographs may, for aught we know, truly represent the parent swarms; it is impossible to imagine that the minor swarm would exactly pass through all the intricacies of those magnificent spirals, and go and return through it precisely on the same path. It would be more probable that, in consequence of perturbations, the secondary swarm would sometimes go through a denser portion, at other times through a less dense portion, and that would be quite sufficient to give us a considerable difference of luminosity.

That the intervals between successive maxima and the mag-

---

[1] *General Astronomy*, 1891, p. 485.

nitudes at maxima are not constant does not prove that the actions are not periodic.

The only objections which have yet been made to the new view as to the cause of variability in this class of stars are thus readily explained.

But suppose for a moment that this view of two bodies is not accepted. What have we got in place of it? We have to explain all the phenomena of variability by one body. That has been attempted more or less happily. Suppose, for instance we have the case of a body waxing and waning quite regularly; we have only to say that body is like a soup-plate, and rotates on an axis, so that sometimes we see the face, sometimes only the edge. But that is not very satisfactory, because we do not know any stars which are like soup-plates. Another way is to say that the stars which are variable have great dark patches on one side of them, great bright patches on the other. We can get a variation of light by such a scheme as that; but we have not observed that; we are simply inventing, merely suggesting ideas to nature that I fancy nature will tell us by-and-by are quite erroneous. For instance, I have stated the facts with regard to β Lyræ which establish the action of two bodies, and I have given the explanation formerly put forward for the variability of that star. The old explanation is blown into thin air by the facts. I think it will be acknowledged that the old ideas are irrational, because they have no true basis of fact, and we must remember that in all this work we must deal strictly with the facts in accordance with the rules of philosophising; *i.e.*, we must never have a complicated explanation until we are perfectly certain that a simpler explanation will not do, and the simplest explanation of all is that which occurs most frequently in the region of facts. That puts the soup-plate theory with regard to variable stars entirely out of court. Further we must remember that, supposing those who still hold to the one-body theory, one star, one variability, object to the possible explanation of variability by the meteoritic hypothesis, they will find it very much more difficult to explain the departure from regularity by any geometric system, because a geometric system must certainly be more rigid than any other, and, therefore, any irregularity under it would be almost impossible.

Lastly, by way of reminder, it may be stated that[1] any amount of variability and *irregular* variability in the light curves of these bodies may be explained on very simple grounds, provided we acknowledge that we are dealing with the move-

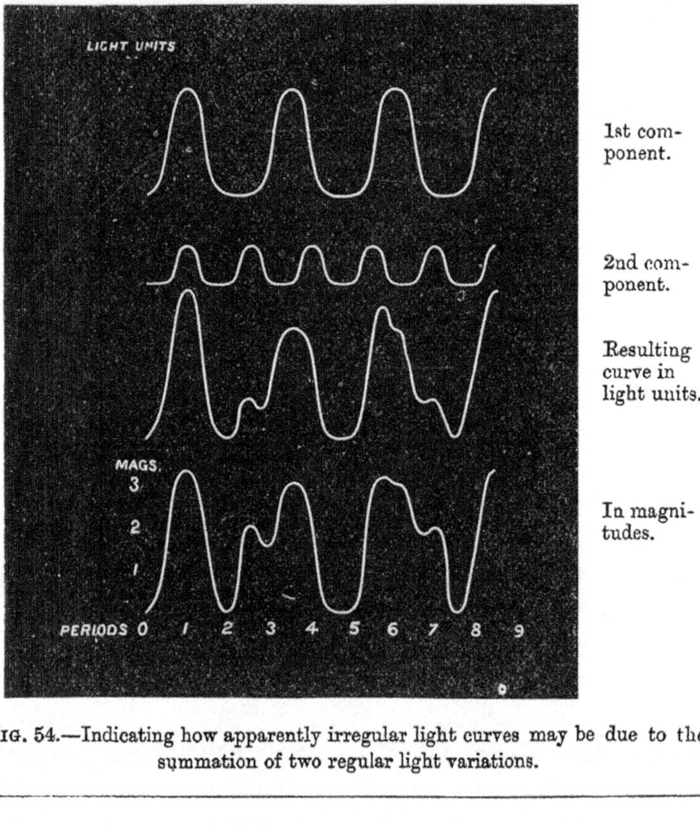

FIG. 54.—Indicating how apparently irregular light curves may be due to the summation of two regular light variations.

[1] *Meteoritic Hypothesis*, chap. xliv.

ments of more than two bodies. For instance, suppose we have one cause at work which gives us a maximum and minimum, and another cause which gives us two very much smaller maxima and minima occurring at a different period represented in Fig. 54 in the upper part of the diagram.

If we add these two together, we get the irregular light curve shown below the two simple curves in the diagram. But the amount of irregularity may possibly only reveal the amount of our ignorance, and when the time comes when we can isolate these two causes, and thus see how the addition of them should be made, we shall find that every part of this curve is really the result of a most beautiful law. I am very glad to say that quite recently Mr. Maxwell Read, of the Harvard Observatory, has put forward the same suggestion, so that we may hope that it will soon be worked out on pretty broad lines.

CHAPTER XI.—EYE OBSERVATIONS OF NEW STARS.

Early Views.

Closely associated with variable stars are the phenomena of so-called "new stars." We have had during the last thirty years five of these new stars, and it was during the appearance of one in the constellation Cygnus in 1876 that I was led to the views which I still hold with regard to their origin.

One of the most remarkable features of these new stars is the rapidity with which they lose their brilliancy.

The first published views relating to the possible causes of the phenomena presented by stars suddenly coming into visibility we owe to Tycho Brahe and Kepler. These were promulgated with reference to, and in explanation of, the new stars which appeared in 1572 and 1604. It will be convenient to begin our consideration of the views in question by giving a condensed account of the observations which had suggested them.

The Nova of 1572 observed by Tycho Brahe.

This, the first Nova of which anything like a complete record exists, appeared in Cassiopeia, in 1572. The accompanying phenomena were minutely described by Tycho Brahe,[1] who first saw it on November 11, 1572. The Nova appeared destitute of nebulous surroundings, and only differed from

[1] Tycho Brahe, *Progymnasmata*, lib. ii, p. 300.

other stars in the vivacity of its scintillations. When it was first observed, it appeared more brilliant than Sirius, α Lyræ, or Jupiter, and even rivalled the splendour of Venus at greatest brilliancy, being, like Venus, visible in the daytime. At the beginning of December a diminution in brightness was noticed. This regularly continued until, in March, 1574, the Nova had disappeared. The gradual diminution of the star's luminosity may be seen from the following photometric observations made by Tycho Brahe :—

November, 1572. Brighter than Sirius, α Lyræ, or Jupiter, and only comparable to Venus when most brilliant.

December, 1572. As bright as Jupiter.

January, 1573. Less bright than Jupiter.

February and March, 1573. Equal to a first magnitude star.

April and May, 1573. Equal to a second magnitude star.

July and August, 1573. Equal to a third magnitude star.

October and November, 1573. Equal to a fourth magnitude star.

End of November, 1573. Equal to a fifth magnitude star.

Between December, 1573, and February, 1574. Equal to a sixth magnitude star.

March, 1574. Disappeared.

Changes of colour accompanied the changes of brightness. When the star first became conspicuously visible it was white, like Venus and Jupiter. It then acquired a yellow colour which merged into red. In the first months of 1573, Tycho Brahe compared it to Mars and α Orionis, and considered it to be much like Aldebaran. Later on in the same year, and especially towards May, a leaden hue was observed, like that exhibited by the planet Saturn. This continued until January, 1574, when the colour appeared to become less clear and of a less pure whiteness as the star slowly disappeared.

The Nova of 1604 *observed by Kepler.*

The famous Nova which appeared in 1604 is associated with the name of Kepler, as that of 1572 is with Tycho Brahe. It was first observed on October 10, by Bronowski, a pupil of Kepler's. The latter astronomer saw the Nova on October 17, and wrote a circumstantial account of his subsequent observations.

The following is a tabulated account of the estimations of its magnitude during the fifteen months that the star was visible:—

1604, October. Brighter than first magnitude stars, and also Saturn, Mars, and Jupiter.

1604, November 6. Could be seen at twilight, although Jupiter was not visible.

1604–5, December 24 to January 3. Greater than Antares, but less than Arcturus.

1605, March 20. Less than Saturn, but greater than third magnitude stars.

1605, April 21. Equal to a third magnitude star.

1605, August 12 and 14. Equal to a fourth magnitude star.

1605, September 13. Less than a fourth magnitude star.

1606, March. Disappeared.

Kepler does not record a sequence of changes in colour like that undergone by the Nova of 1572, although he alludes to the yellow, saffron, purple, and red tints of the star when it was near the horizon. It will be seen from the following extract[1] that he refers to the scintillations as being like the colours reflected by the faces of a diamond, but this was also the case with Tycho Brahe's Nova, and is distinct from the gradual colour changes of the latter star.

---

[1] *De Stella Nova in Pede Serpentarii*, Prague, 1606, p. 5.

"Mihi, qui potissimum respiciebam ad colorum Iridis varietatem, quos pro scintillationis alio atque alio habitu evomuit, placuit uti exemplo Adamantis multanguli, qui Solis radios inter convertendum, ad spectantium oculos, variabili fulgore revibraret: simul ad causam scintillationis, in Opticis meis allatam, alludere volui.

"De coloribus Mæstlinus planè mecum, eos ad momenta singula variari, ex flava mox croceam, è vestigio purpuream et rufam, *ut plurimum candidam videri, ubi ex vaporibus paulo altius elevetur.*"

Although many other Novæ have been observed, none have matched the splendour of those of 1572 and 1604, and of none have such circumstantial accounts been written. I now proceed to a statement of the explanation of the phenomena put forward by the respective observers.

Tycho Brahe's Hypothesis.

Tycho Brahe considered that new stars were formed from the cosmical vapour which was supposed to have reached a certain degree of condensation in the Milky Way.[1]

"Cœli materia ut subtillissima nostroque visui et planetarum circuitibus pervia, in unum tamen globum condensata compactaque, et lumine, si non proprio saltem solari, illustrata, hanc stellam effingere potuit. Quæ quoniam citra communem naturæ ordinem quasi monstrosa extitit, parem cum cæteris perseverantiam obtinere nequibat; veluti neque novæ ex elementis constantes generationes et monstra diu durant.

"Et quamvis in tota cælestis mundi vastitate materia pro conformatione alicuius stellæ ascititiæ, meo judicio abundè suppetat; tamen nusquam copiosius et pleniùs, quam juxta Viam Lacteam, quam substantiam quandam cælestem à materia reliquarum stellarum non discrepantem, sed diffusam certisque locis expansam, non in unum corpus discretim, prout in stellis fit, conglobatam esse statuo: hincque factum judico quod nova hæc in ipso Galaxiæ margine constiterit."

It will be seen from the above extract that the fact that the Nova appeared on the edge of the galaxy was used by Tycho Brahe to give weight to the hypothesis of stellar formation advanced by him. Indeed, some observers imagined they could

[1] *Progymnasmata*, p. 794.

see the *hiatus* or opening out of which the Nova came. The disappearance of the star was supposed to be due either to some action in itself or to its dissipation by the light of the sun and stars. When Tycho Brahe advanced the above theory, the tails of comets were looked upon as similar in constitution to the Milky Way.

Support of the Hypothesis by Kepler.

Kepler agreed with Tycho in considering that new stars were created from the ethereal substance of which the Milky Way was composed. The circumstance that Mira Ceti, which was looked upon as a Nova, appeared in a part of the heavens distant from the Milky Way, was explained by saying that the nebulous material was not exclusively confined to the galaxy, as supposed by Tycho Brahe, but pervaded all space.[1]

A fact deemed of considerable importance was that both Tycho Brahe's and Kepler's Novæ became suddenly and strikingly visible, and did not appear gradually to increase in brightness. Indeed, it was thought that all new stars must exhibit the maximum of brilliancy at their first appearance, and Kepler went so far as to use the statement made by Antonius Laurentinus Politianus, that he had seen the Nova of 1604 increase in brightness, as an argument against his having seen the star at all. The following are the words used by Laurentinus :—"Apparuit parva, et postea de die in diem crescendo apparuit magnitudine, et lumine non multò inferior Venere, superior Jove."[2]

Newton's View.

The first Nova that attained any brilliancy, after that of 1604, appeared near β Cygni in June, 1669, and was observed by

---

[1] *De Stella Nova in Pede Serpentarii*, Kepler, pp. 110–112.
[2] *Ibid.*, p. 3.

Anthelm.[1] This Nova fluctuated in brightness between the 3rd and 5th magnitudes, and finally disappeared altogether. It is most probable that observations of this star drew Newton's attention to the subject, and led him to the idea that "Novæ" were produced by the appulse of comets, thus propounded in 1686 in the *Principia*.[2]

"Sic etiam stellæ fixæ, quæ paulatim expirant in lucem et vapores, cometis in ipsas incidentibus refici possunt, et novo alimento accensæ pro stellis novis haberi. Hujus generis sunt stellæ fixæ quæ subito apparent, et sub initio quam maxime splendent, et subinde paulatim evanescunt. Talis fuit stella in cathedra Cassiopeiæ quam Cornelius Gemma octavo Novembris, 1572, lustrando illam cœli partem nocte serena minime vidit; at nocte proxima (Novem. 9) vidit fixis omnibus splendidiorem et luce sua vix cedentem Veneri. Hanc Tycho Brahæus vidit undecimo ejusdem mensis ubi maximè splenduit; et ex eo tempore paulatim decrescentem et spatio mensium sexdecim evanescentem observavit."

More Modern Views.

In dealing with the period between Newton's time and 1890, I propose to give, as shortly as possible, some of the most important views expressed, during the last quarter of a century, by observers of the phenomena we are now considering.

Zöllner, 1865.

According to the hypothesis advanced by Zöllner,[3] all stars, at a certain period of their formation, become covered with a cold, non-luminous crust. If the glowing mass bursts forth, the chemical combinations which have formed on the surface, under the influence of a low temperature, are again decomposed with a resulting development of considerable heat and light. Hence the great brilliancy of a new star must not be ascribed

[1] *Phil. Trans.*, vol. 5, p. 2028.
[2] P. 525, Glasgow edition, 1871.
[3] *Photometrische Untersuchungen*, 1865, p. 251.

merely to the bursting forth of a glowing mass, but also to the combustion of the substances which form the shell.

Huggins and Miller, 1866.

Drs. Huggins and Miller's observations of the Nova that appeared in Corona Borealis in 1866 led them to the following general conclusions :[1]

"It is difficult to imagine the present physical condition of this remarkable object. There must be a photosphere of matter in the solid or liquid state emitting light of all refrangibilities. Surrounding this must exist also an atmosphere of cooler vapours, which give rise, by absorption, to the groups of dark lines.

"Besides this constitution, which it possesses in common with the sun and the stars, there must exist the source of the gaseous spectrum. That this is not produced by the faint nebulosity seen about the star is evident by the brightness of the lines, and the circumstance that they do not extend in the instrument beyond the boundaries of the continuous spectrum. The gaseous mass from which this light emanates must be at a much higher temperature than the photosphere of the star, otherwise it would appear impossible to explain the great brilliancy of the lines compared with the corresponding parts of the continuous spectrum of the photosphere. The positions of two of the bright lines suggest that this gas may consist chiefly of hydrogen. . . . . The character of the spectrum of this star, taken together with its sudden outburst in brilliancy and its rapid decline in brightness, suggests to us the rather bold speculation that, in consequence of some vast convulsion taking place in this object, large quantities of gas have been evolved from it, that the hydrogen present is burning by combination with some other element and furnishes the light represented by the bright lines, also that the flaming gas has heated to vivid incandescence the solid matter of the photosphere. As the hydrogen becomes exhausted all the phenomena diminish in intensity and the star rapidly wanes."

Johnstone-Stoney, 1868.

In a paper on the physical constitution of the sun and stars, Mr. Johnstone-Stoney, in 1868, discussed the origin of double stars. He then remarked[2] that the astonishing phenomena

---

[1] *Proc. Roy. Soc.*, vol. xv, p. 148.
[2] *Ibid.*, vol. xvii, p. 53.

witnessed in T Coronæ were precisely what should arise towards the end of a process he describes; this process may be gathered from the following quotation :—

"The stars having been intensely heated by previous perihelion passages, and having begun to shrink, would, at ordinary times, present a spectrum subdued by an abundance of very dark lines; but immediately after one of the last occasions upon which their atmospheres brush against one another, the outer constituent of their atmospheres [hydrogen] and the outer constituent alone, would be raised by the friction to brilliant incandescence, which would reveal itself by the temporary substitution of four intensely bright for four dark hydrogen lines in a spectrum which everywhere else continues to be filled with dark lines. And, moreover, these dark lines would for a while be rendered faint by the fierce heat radiated upon the outer parts of the atmosphere of each star by its companion."

Vogel, 1877.

Observations of the new star in Cygnus (1876—77) led Professor Vogel to remark as follows :—[1]

"It is generally supposed that the bright lines in some star spectra are due to gases which burst out from the interior of the luminous body at a temperature higher than the surface, as similar lines are occasionally seen in the spectra of sun-spots, and are caused by the eruption of incandescent hydrogen from the hot interior over the cooler surface of the spot. But this is not the only explanation. It may be assumed that the outer shell of a star, consisting of glowing gases, as in the case of our sun, has a lower temperature than the nucleus, although it is at a relatively high temperature.

"On the first supposition, it is impossible for the appearance to last for any length of time. The gas bursting forth from the heated interior would communicate part of its heat to the cooler surface, thus raising the temperature of the latter until the difference of temperature between the incandescent gas and the surface is insufficient to produce the bright lines, which consequently disappear.

"This theory satisfactorily explains the apparition of so-called *new* stars, having bright lines in their spectra, and which suddenly appear and speedily disappear altogether or lose most of their brilliancy, if Zöllner's hypothesis be accepted as a further explanation."

---

[1] *Berlin Akad. Monatsberichte*, 1877, p. 256.

Lohse, 1877.

Dr. Lohse advanced a theory to account for "new stars," founded upon his observations of Nova Cygni.[1] His suggestions are summed up as follows :—

(1) "The lighting up of new stars may probably be looked upon as the result of the innate affinity of chemical matter.

(2) "The conditions and manner of illumination may be regarded as follows :—By the progressive cooling of the mass of a luminous body (fixed star) which consists of heated vapours and gases, an atmospheric envelope is erected which absorbs the light so much, that the star cannot be seen at all, or only very faintly, from the earth. As this body continues to give out heat, at length the degree of coolness is reached which is necessary for the formation of chemical combinations. The greater portion of the body is composed of elements which then combine, producing, by their combination, heat and light, and thus making the star visible to a great distance, and for a long or short space of time."

Lockyer, 1877.

In this year, discussing the phenomena of Nova Cygni, I advanced the view that meteoritic collisions were, in all probability, the cause of them. This will be fully discussed later on.

Monck, 1885.

Mr. W. H. S. Monck suggested, in relation to the new star in Andromeda, that :—

"As shooting-stars are known to be dark bodies rendered luminous for a short time by rushing through our atmosphere, new stars are dark (or faintly luminous) bodies which acquire a short-lived brilliancy by rushing through some of the gaseous masses which exist in space."

---

[1] *Berlin Akad. Monatsberichte*, p. 826.

CHAPTER XII.—THE METEORITIC THEORY OF THE ORIGIN OF NOVÆ.

THE whole conception of the meteoritic hypothesis arose from a consideration of those bodies which sometimes quite suddenly make their appearance in the heavens.

It was a "new star" which appeared in 1876 which drew my special attention to these phenomena, and, as a result, I wrote as follows:—

"It should have been perfectly clear to those who thought about such matters, that the word 'star' in such a case is a misnomer, from a scientific point of view, although no word would be better to describe it in its popular aspect. The word is a misnomer, for this reason. If any star, properly so called, were to become a 'world on fire,' were to 'burst into flames,' or, in less poetical language, were to be driven either into a condition of incandescence absolutely, or to have its incandescence increased, there can be little doubt thousands or millions of years would be necessary for the reduction of its light to the original intensity. Mr. Croll has recently shown that if the incandescence observed came for instance from the collision of two stars, each of them half the mass of the sun, moving directly towards each other with a velocity of 476 miles per second, light and heat would be produced which would cover the present rate of the sun's radiation for a period of 50,000,000 years. A very different state of affairs this from that which must have taken place in any of the Novæ from the time of Tycho to our own, and the more extreme the difference, the less can we be having to deal with anything like a star properly so called.

"The very rapid reduction of light in the case of the new star in Cygnus was so striking that I at once wrote to Mr. Hind to ask if any change of place was observable, because it seemed obvious that if the body which thus put on so suddenly the chromospheric spectrum were single, *it might only weigh a few tons or even hundredweights*, and being so small might be very near us. We are driven, then, from the idea that these phenomena are produced by the incandescence of large masses of matter because, if they were so produced, the running down of brilliancy would be exceedingly slow.

"Let us consider the case, then, on the supposition of small masses of matter. Where are we to find them? The answer is easy: in those small meteoric masses which an ever-increasing mass of evidence tends to show, occupy all the realms of space.

It may here be stated that among the well-observed phenomena recorded, the star on dimming its lustre gave out light spectroscopically identical with that emitted by the nebulæ. I pointed out that this fact

"Not only goes far to support the view I have suggested as against that of Zöllner, but it affords collateral evidence of the truth of Thomson and Tait's hypothesis of the true nature of nebulæ. The nebular hypothesis in its grandeur and simplicity remains untouched by these observations, the facts, so far from being in direct opposition to it, help us, I think, all the better to know exactly what a nebula is.

"There is another point of extreme interest to the spectroscopist, if we accept the bright line observed in the star by Dr. Copeland and others to be veritably the chief nebula line. It is clear from Dr. Vogel's diagram, that this line brightened *relatively* with each decrease in the brilliancy of the hydrogen lines. On December 8th, 1876, it was much fainter than F, while by March 2nd, 1877, F was a mere ghost by the side of it. On any probable supposition the temperature must have been higher at the former date."[1]

Returning to the subject of new stars in 1887, in a preliminary discussion of the meteoritic hypothesis, I saw no reason to change my views, and an inquiry into the spectroscopic phenomena led me to state that:—

"New stars, whether seen in connection with nebulæ or not, are produced by the clash of meteor swarms, the bright lines seen being low temperature lines of elements, the spectra of which are most brilliant at a low stage of heat."[2]

This general theory, however, does not exclude other causes, such as the sudden illumination produced by collisions of two

---

[1] *Nature*, vol. xvi, p. 413, 1877.
[2] November, 1887, p. 154.

dark bodies; but I pointed out that no case of this kind has happened within human ken.

It follows, then, that the phenomena of new stars are produced by the same cause as that which is at work in the variable stars in which we get the greater light produced at the moment when two swarms, one revolving round the other, are nearest together. The only difference being that, in the variable stars, we deal with periodic phenomena.

It is obvious, then, that a complete discussion of these phenomena should afford a valuable test of the general hypothesis, for the reason that such bodies, instead of going forward along the temperature curve, should go back as they cooled, and the sequence of spectra should be reversed.

In 1890 all the observations, therefore, were brought together and discussed from this point of view. These discussions form the subject matter of the present chapter, which consists of a condensation and slight revision of a paper I communicated to the Royal Society in that year.[1] The investigation had particular reference to the sequence of the spectra of Novæ from their appearance to their diminution to invisibility. In the absence of spectroscopic observations the changes of colour were studied to afford a means of arriving at some idea of the physical constitution of the body observed.

For the purposes of this inquiry I define a Nova as a body which suddenly appears, then diminishes its brightness, and, finally, disappears *as a star*.

Discussion of the Spectra of Novæ.

Nova Coronæ, 1866.

The spectrum of this star was observed on several dates by Messrs. Huggins and Miller, Wolf and Rayet, and Stone and

[1] *Phil. Trans.*, 1891, A, p. 397, *et seq.*, November 28, 1890.

Carpenter. The observations of Huggins and Miller were made on May 16, 17, 18, 19, 21, and 23; those of Wolf and Rayet on May 20; and those of Stone and Carpenter on May 19, 20, 23, 24, 28, and June 7. There is a general agreement as to the positions of the lines determined by the different observers, and the principal difference observed, as the star faded, was simply a diminution of the number of lines.

In their first observation, on May 16, Messrs. Huggins and Miller[1] noted four bright lines, and suspected a fifth. Two of these were found by direct comparison to be the C and F lines of hydrogen, but the positions of the other two were only roughly plotted on a diagram. There are not sufficient data for accurate reduction of the wave-lengths of these lines, but by a curve they have been found to be at approximately 468 and 473. Of the more refrangible of the two it is remarked: "The appearance of this line suggested that it was either double or undefined at the edges." The fifth line suspected was probably not far from G of the solar spectrum, and the presence of C and F make it highly probable that this was the hydrogen line at G.

In addition to these bright lines, many absorption lines and bands were observed. These absorptions were evidently very similar to those seen in stars like α Orionis. This was noted by Messrs. Huggins and Miller, and I have since compared those in the blue end of the spectrum with the lines photographed in α Orionis by Professor Pickering, who kindly furnished me with the data necessary for the reduction to wave-lengths. This shows a very close relation between the two spectra. The spectrum was again observed by Messrs. Huggins and Miller on May 17 and 19, but no remarkable difference was noted.

On May 19, the spectrum was also observed by Messrs. Stone

[1] *Proc. Roy. Soc.*, vol. xv, p. 147.

and Carpenter.[1] Four bright lines were seen, but only three of them were fixed by measurements. The position of the fourth was approximately determined by estimation. One of the lines was undoubtedly F, and the wave-lengths of the others, as deduced from a curve, I find to be about 501, 471, 467. The data for reduction, however, are quite insufficient to give anything like accurate results. The C line was not recorded.

In the account of his observations of Nova Cygni, Vogel refers to the observations[2] of Stone and Carpenter; he reduced the positions of the lines to wave-lengths 502, 467, 463. He points out, however, that on account of insufficient data, these cannot be regarded as accurate.

The line 500 was seen only by Messrs. Stone and Carpenter. The other two lines are, in all probability, identical with those observed by Messrs. Huggins and Miller, for which I have determined the wave-lengths 468 and 473. Traces of absorption lines were also observed, thus confirming the observation of Messrs Huggins and Miller. Another observation was made by Messrs. Stone and Carpenter on May 20, but to them the spectrum did not differ from that of the previous day.

MM. Wolf and Rayet,[3] however, made an observation of the spectrum of this star on the same day. They recorded :—

"The light of the new star gives a complete, very pale spectrum, on which are seen a certain number of brilliant bands. The brightest and largest of these bands appeared in a continued manner almost at the edge of the yellow and the green. It is preceded on the yellow side by a rather dark space, then by a bright, but weak line. A third line, which seems to correspond to D, is seen in the yellow, pretty bright, and towards the orange. Then, after the brightest line towards the violet end, the green is seen well marked, then a darker space a little darker than we spoke of before, and another line as bright as the principal band. The rest of the spectrum is pale, with ill-defined edges, and with nothing marked about it that we could distinguish."

---

[1] *R. Astron. Soc. Monthly Notices*, vol. xxvi, p. 295.
[2] *Berlin Akad. Monatsber.*, 1877, p. 243.
[3] *Comptes Rendus*, vol. lxii, p. 1108.

It is greatly to be regretted that the positions of the lines were not measured. Without these it is quite impossible to connect them with the other observations. It is remarkable that the line in the yellow, which may have been D or D^3, was not recorded by any of the other observers.

Messrs. Huggins and Miller again observed the spectrum on May 21 and 23. On these dates the bright lines appeared more brilliant than on former occasions, owing to the dimming of the continuous spectrum. On May 23 Mr. Stone again saw the four lines which he had observed previously, but on May 24 and 28 and June 7 he was only able to see the F line. It was the opinion of both Mr. Stone and Mr. Carpenter that the bright lines and the continuous spectrum faded at the same rate.

All these observations, with the exception of those by Wolf and Rayet, are shown on the accompanying map.

There can be no question as to the origin of two of the lines—that in the red and that in the blue-green. They were undoubtedly due to hydrogen. The line near 501 was, in all probability, the same as that which was seen to brighten in Nova Cygni as the star faded. This was also suggested to Vogel in referring to his observations of Nova Cygni. There is other evidence to show that this was identical with the chief line in the spectrum of the nebulæ.

The two lines in the blue present a little more difficulty. The more refrangible one near 468 may have been due to the carbon fluting, which, under certain conditions, has its maximum intensity about 468,[1] especially as Dr. Huggins stated that its "appearance suggested that it was either double or undefined." The one near 473 was probably the less refrangible edge of the same compound fluting of carbon.

In some of the observations of comets (*e.g.*, Comet III, 1881,

---

[1] *Proc. Roy. Soc.*, vol. xlv, p. 167.

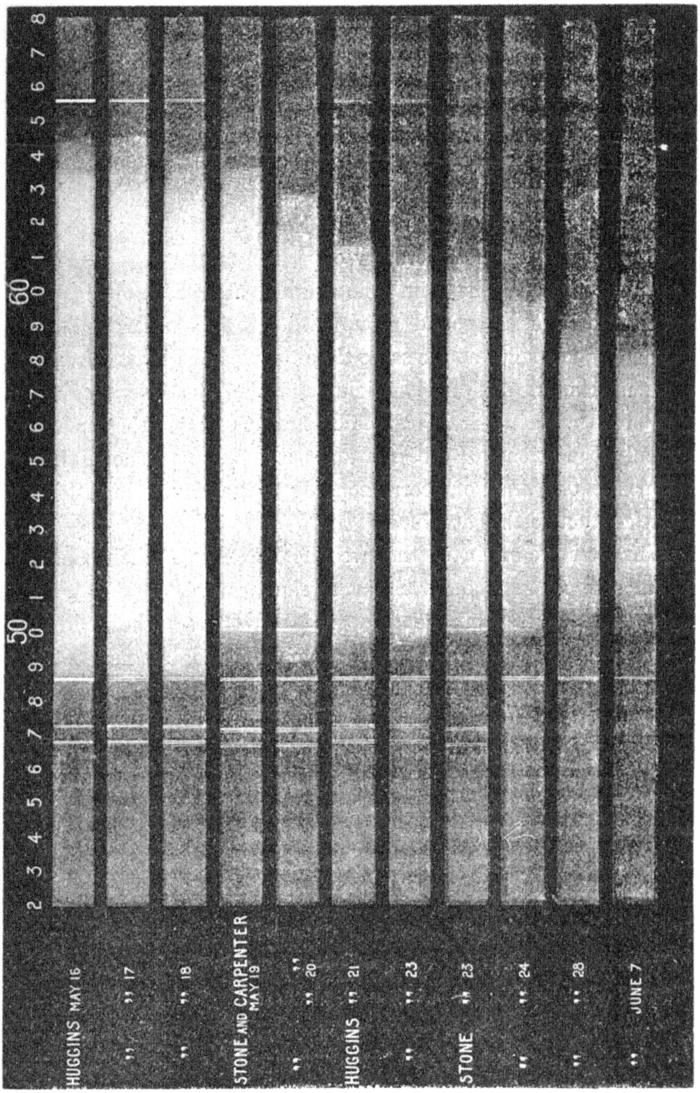

FIG. 55.—Spectra of Nova Coronæ.

on June 28) the blue band has been recorded with two maxima, one at 468 and the other at 474. This strengthens the view that the two lines seen in Nova Coronæ may have had the same origin. In comets, the more refrangible one is the least defined, and this was also the case in the Nova. The two maxima were also seen subsequently in Nova Cygni, but in that case the less refrangible one was only visible for a short time. I have very little doubt that both in comets and Novæ the two maxima are due to carbon. In the laboratory experiments (see *ante*, p. 149) in connection with this point photographs have been obtained, which show the compound group with a bright maximum of intensity about 468, at the same time leaving the least refrangible maximum (474) fairly bright, and the two intermediate ones much weaker. It is not difficult to understand that under difficult conditions of observation such a group should be mapped as two lines, one corresponding to the maximum at 468 and the other to that at 474.

Whatever the origin of the two lines, the fact of their being common to comets and Novæ is very significant.

It is rather remarkable that in Nova Coronæ the F line should remain visible for a longer time than the line near 500, because in Nova Cygni it died away first, leaving 500 very bright to the end. It must be remembered, however, that the later observations by Mr. Stone were made under very difficult conditions. The 500 line would also be more liable to be masked by continuous spectrum than the F line, so that the presence of a brighter continuous spectrum than in Nova Cygni may explain why F should be seen last in one case and 500 in the other.

I have already pointed out that in Novæ we have not simply to deal with one state of condensation of a swarm. At least two swarms are concerned, and these are not necessarily of the same density. It therefore becomes interesting to attempt to reproduce the spectrum of a Nova by integrating the spectra of two or more swarms of meteorites, each of which is at a stage of condensation different from the others. Such an attempt is shown in Fig. 56, where the spectra of Coggia's comet and the planetary nebula G.C. 4373 are integrated. The comet at this stage showed the carbon spectrum, the blue band having two maxima. In the resulting integration the green cometary bands are almost masked by continuous spectrum, whilst the blue one remains visible, exactly as it does in the bright-line stars.[1] The nebula line 495 was probably too faint to be detected. The line 500 was also faint, and this and 495 were probably partly masked by the continuous spectrum, which would be slightly brighter in that part of the spectrum than at F. The continuous spectrum would be brightened by the collision of two swarms, and hence it is necessary to add continuous spectrum in the integration.

The magnitude of this Nova varied from 2 on May to 9 on June 7.

Nova Cygni, 1876–77.

This star was discovered by Schmidt at Athens on November 24, 1876. Its magnitude was then 3·0, and its colour a reddish-yellow. On December 5 the magnitude was 5·9, and it steadily diminished in brightness, until on March 10 its magnitude was 8·3. In 1882 Dr. Copeland found that it had decreased to the 14th magnitude.

The spectrum was observed on several occasions by Cornu, Vogel, and Copeland. The earliest observations appear to have been made by M. Cornu, who recognised bright lines on December 2, but was unable to make any measurements. Two days after, a more complete examination of the spectrum was made, and on this occasion the positions of the lines were measured. Eight lines were bright enough to be recorded, but no dark

---

[1] *Proc. Roy. Soc.*, vol. xliv, p. 33.

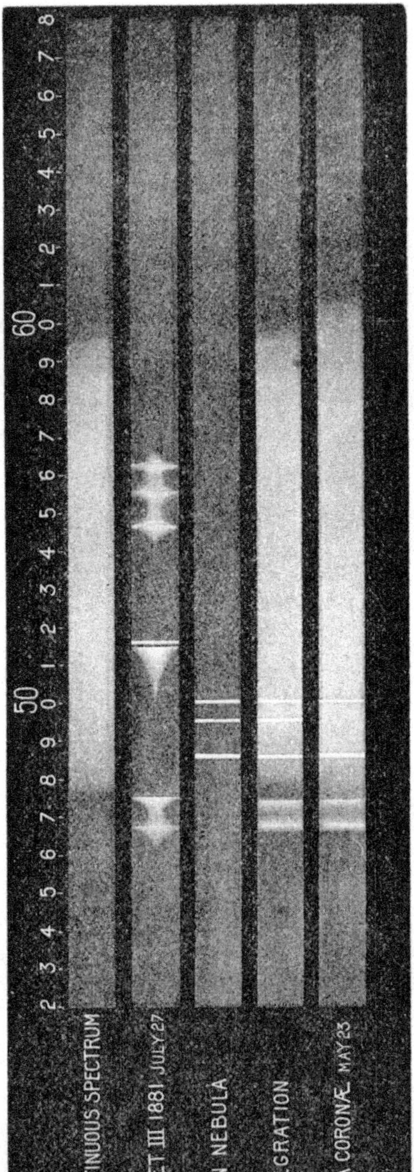

Fig. 56.—Integration of the spectra of Coggia's comet and planetary nebula compared with spectrum of Nova Coronæ.

lines were seen. He says,[1] "The dark lines, if they exist, must be very fine, and must have escaped me, on account of the very feeble light of the star." The following are the wave-lengths of the bright lines observed, the Greek letters indicating the relative intensities, a being the brightest :—

α.	δ.	γ.	β.	ζ.	η.	θ.	ϵ.
661	588	531	517	500	483	451	435

The lines a, δ, η, and ϵ were doubtless C, D, F, and G respectively, and M. Cornu points out that the remaining lines, with the exception of the one near λ 500, are nearly coincident with the lines which occur most frequently in the spectrum of the solar chromosphere. The line γ is regarded by M. Cornu as possibly coincident with the coronal line 5315·9. β he believed to correspond with the b lines of magnesium. He further states that—

"The feeble line θ corresponds also to a line, $\lambda = 447$, of the chromosphere; one is thus led to think that the line δ corresponds rather to the bright line of the chromosphere, $\lambda = 587$ (helium), than to that of sodium, 589. If this interpretation be accurate, the bright lines of the spectrum of the star comprehend inclusively the brightest and most frequent lines of the chromosphere."

In the diagram of the spectrum which accompanies M. Cornu's paper, there are shown three maxima of brightness which are not referred to in his description. These have been reduced by a curve, and their wave-lengths found to be approximately 546, 563, and 635. These agree very well with lines observed by Vogel.

Vogel thus summarises the lines which he observed,

(1) The hydrogen lines, Hα } certainly.
Hβ }
Hγ, most probably.

(2) A line of the wave-length 499 $\pm$ 1 max. This line comes into pretty close agreement with the brightest line of

[1] *Comptes Rendus*, vol. lxxxiii, p. 172.

the nitrogen spectrum under ordinary pressure; it is the same line which is brightest in the spectrum of the nebulæ.

(3) A faint line at λ 580.

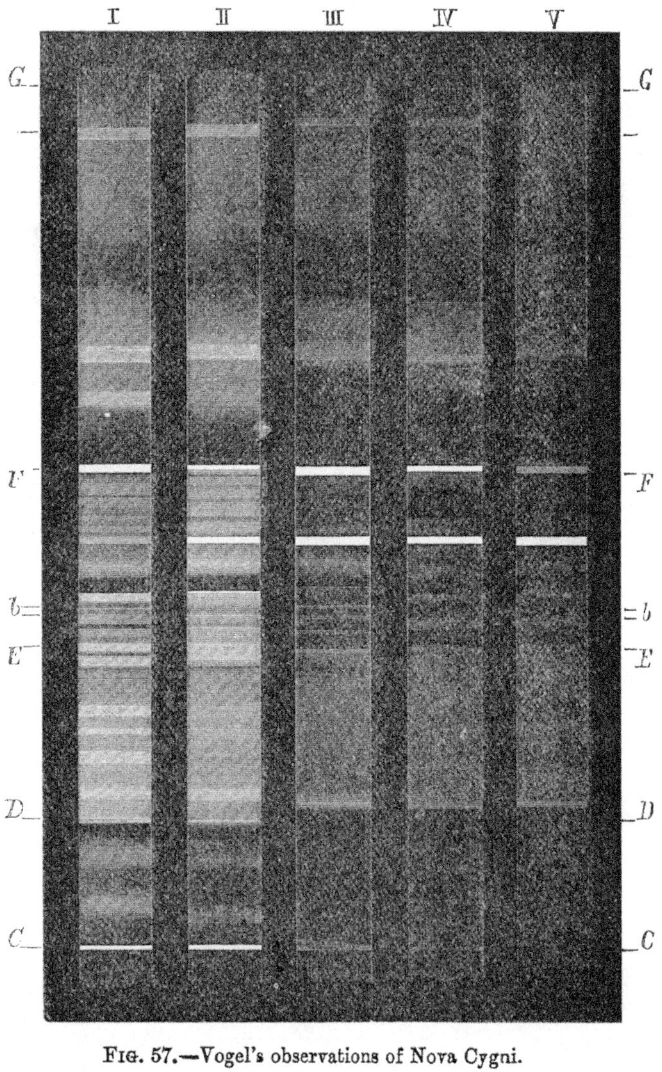

FIG. 57.—Vogel's observations of Nova Cygni.

(4) A similar one at 467. (This agrees pretty well with a group of lines close together in the spectrum of air.)

(5) There were also bright lines seen in the region of *b* and E, but nothing certain could be ascertained about their position. At the first observation on December 5, two lines were measured in the blue (474 and 470); these were again seen on December 8, but in later observations only the second was perceived as a faint band (λ 467).

For further information the maps must be referred to, for, as Vogel states, they show several details which can hardly be put into words.

Vogel refers to Cornu's observation, and states that he cannot agree that "the atmosphere of the star possesses exactly the same composition as that of the chromosphere of the sun." His chief objection to this view is that the line near 500, which is not in the chromosphere, was distinctly seen along with other bright lines in the star's spectrum, and eventually became the strongest line.

Dr. Copeland[1] also made a large series of observations, but he was not able to commence before January 2, 1877. He continued his observations to February 16, and recommenced on September 2.

On January 2 five bright lines were recorded. The measurements show that two of these lines were the C and F lines of hydrogen. Two series of measurements were made, and for the other three lines the following wave-lengths were determined:—

1st series.	2nd series.	Notes.
577·9	579·0	Bright band fading rapidly on both sides.
502·4	505·1	Bright well-defined lines.
463·4	466·8	Faint band.

[1] *Copernicus*, vol. ii, p. 102.

There can be little doubt that these correspond to the lines measured by Vogel at 580, 499, and 470 (some measures 467).

On January 9 seven lines were measured, and it was further stated that "the space between the red line (C) and the next one (D) was certainly columnar, and probably contained two maxima." These were probably identical with those seen by Vogel, and they have therefore been inserted in the map (Fig. 67).

Dr. Copeland[1] gives as the adopted means of his observations lines at λλ 656·2, 589·5, 577·5, 502·3, 486·1, 463·5, and 437·6.

The star was not again observed until September 2, 1877 (the estimated magnitude being 10·5), when the startling fact was at once apparent that the spectrum was restricted practically to a single line. The measurements show that this was the line near 500. On September 6 it was observed that, in addition to the chief line, there were traces of one or two lines on the violet side of the chief line, but very close to it. On this date it was noted that the star did not give a sharp image, but that the "extreme diameter could not be above 2 seconds or 1½ seconds." On October 1 the star was examined with a view to detecting any very faint continuous spectrum, but nothing of the kind was visible. On October 10 Lord Lindsay noted that—

"Since the last measures a decided change is to be seen, as the light is more spread out} The mean brightness is still as before, but is divisible into two lines very close together with a dark gap, and then another very faint line."

Measurements of the three lines gave λλ 499·5, 492·2, 491·8.

The latter two "were too feeble and too near together to be effectively measured separately."

But, regarding this, Copeland remarks that—

"The faint extension of the spectrum of this star towards the violet measured on October 10, 1877, was seen with so much difficulty that it

[1] *Copernicus*, vol. ii, p. 111.

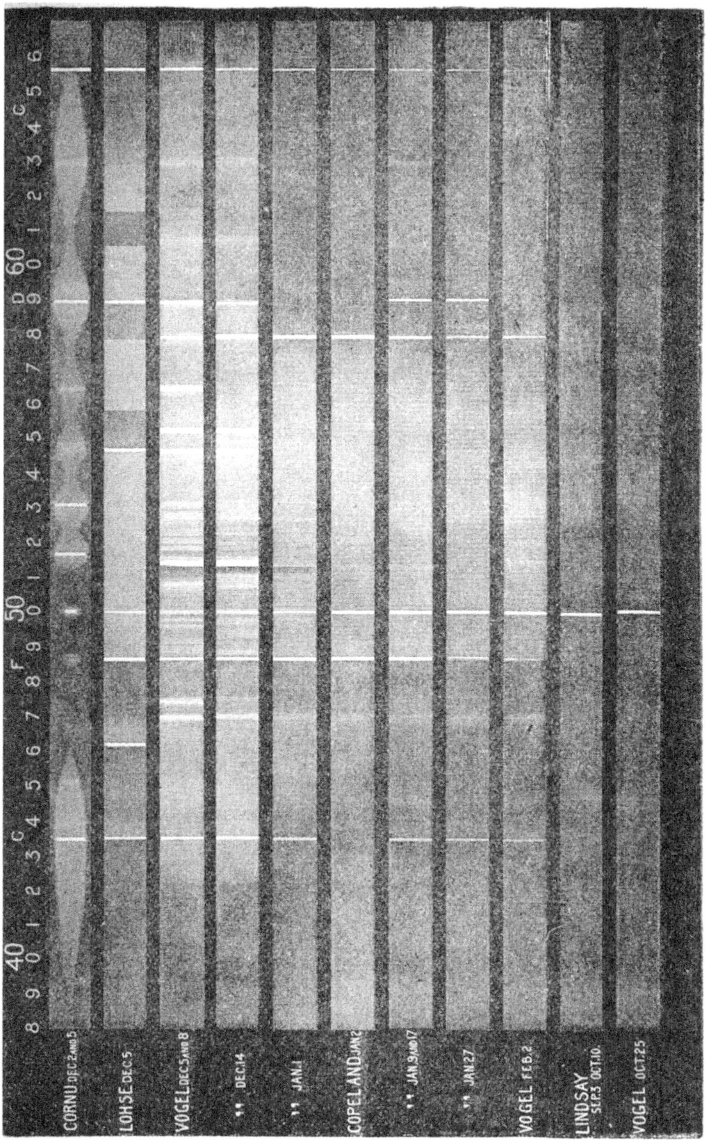

FIG. 58.—Spectra of Nova Cygni.

would be unwise to regard it as in any certain degree indicating the presence of other nebular lines."

On October 25, 1877, Vogel succeeded in making another observation of the star's spectrum.[1] He says:—

"The spectrum was almost monochromatic; it consisted of one bright line, on both sides of which a very faint continuous spectrum could be seen."

This was measured at wave-length 499. This observation was again confirmed on February 18, 1878.

Copeland remarked:—[2]

"Bearing in mind the history of this star from the time of its discovery by Schmidt, it would seem certain that we have an instance before us in which a star has changed into a planetary nebula of small angular diameter,"

and again,[3]

"It seems, therefore, not unreasonable to assume that in the case of a planetary nebula, the whole light of which did not equal the light of a 10th magnitude star, the spectrum might be reduced to a single line, and that this line would then be the one of which the wave-length may be taken at 500·4."

Hence we see that on and after September 3, 1877, the only line seen with certainty was the line near 500, which had gradually increased in brightness since the first observation made on December 5, 1876. All the important spectroscopic observations of Nova Cygni are shown in Fig. 58. It is evident that the hydrogen lines were seen by all the observers, and, notwithstanding the slightly divergent measures, they have all been inserted in their proper places. This also applies to the line near 500. The yellow line we now know must have been D^3, and there is also no doubt about 4472. There is a little more difficulty with the lines in the blue, as Cornu and Lohse differ from Vogel and Copeland. Cornu measured a blue

---

[1] *Berlin Akad. Monatsber.*, 1878, p. 302.
[2] *Ast. Nach.*, No. 2158.
[3] *Copernicus*, vol. ii, p. 113.

line at 451; Lohse's is approximately estimated at 462, while Vogel measured a line at the mean place 467. Copeland measured the line as 463·5, but he is evidently of opinion that Vogel's wave-length is the more correct, on the ground that it agrees with one of the bright bands in the spectra of the bright-line stars in Cygnus. No accurate measures were made of the groups of lines in the green indicated in Vogel's map, and the positions of these have been estimated by reference to those stated to be near 514, 527, and 508.

Secchi also made some observations of the spectrum,[1] and he was of opinion that one of the bright lines coincided with hydrogen, one with magnesium (*b*), and a third with sodium (D). Referring to this, Vogel says somewhat too hastily:—

> "In this he is certainly mistaken, for on January 8 the lines in the neighbourhood of the magnesium group were quite faint, and at D there was not a single bright line to be seen. The bright lines which he noticed had the wave-lengths 500 and 580 mm., and were pretty far distant from the sodium and magnesium lines."

It will be noticed that Secchi's observation agrees to a large extent with that of Cornu, which has been justified by subsequent work.

With regard to the origins of the lines, we have first unquestionably some of the bright lines of hydrogen and of helium. These gradually dimmed as the star diminished in brightness. Again, there was a line near λ 500, apparently coincident with the chief line in the spectra of the nebulæ, which brightened as the other lines faded.

The next in order of importance is, perhaps, the line or band in the blue between F and G, which Vogel at first saw double (470 and 474), and afterwards single, with a mean wave-length of 467. Copeland described it as a "faint band," and gave its mean position as 463·5. It is more than likely that this was the compound fluting of carbon in the blue, seen at first with

[1] *Ast. Nach.*, No. 2116.

two maxima, as in Nova Coronæ and Comet III, 1881, etc., and afterwards with a single maximum at 468. This band agrees in position with the bright band seen in the spectra of bright-line stars (*e.g.*, Wolf and Rayet's stars in Cygnus. See *ante*, p. 149).

Vogel also refers to this relation. The two maxima first seen, however, were not recognised by Vogel as being the same as those observed in Nova Coronæ, although he pointed out that the green line 500 was common to the two as well as the hydrogen lines. In Nova Coronæ there were simply the brightest lines which were afterwards seen in Nova Cygni.

Another important line is the one at 579, about which there is no disagreement between Copeland and Vogel. This is undoubtedly the same line which is seen in the bright line stars in Cygnus, as pointed out by Copeland, and I have previously ascribed it to an iron line which is seen brightest in the flame spectrum.[1]

One of the lines in the green is probably also an iron line, 527, the next in intensity to the line at 579. This died out in the Nova before the line 579, and this is what should happen, for in the laboratory 579 is seen without 527 in the coolest part of the flame. The three lines or bands near 564, 552, and 546, are quite familiar in cometary spectra. The first two of them are probably the 1st and 3rd members of the citron carbon fluting, the intermediate one being masked, as in Coggia's comet on June 13, 1874.[2] The one at 546 is the fluting due to lead, which has frequently been recorded in cometary spectra.

The two bands shown in Vogel's map, at wave-lengths 513 and 517, are probably also carbon bands seen in comets. The less refrangible one is the brighter, exactly as it was in Coggia's comet, on the occasion just referred to.

On December 5, 8, and 14, Vogel recorded four faint lines

---

[1] *Proc. Roy. Soc.*, vol. xliv, p. 33.

[2] *Ibid.*, vol. xlv, p. 175.

between F and the line 500. One of these is doubtless the nebula line 495; the two between this and F are probably identical with two shown in my photograph of the spectrum of the nebula in Orion.[1] The other line has not been identified.

Another line of considerable interest is that observed by Vogel and Cornu, near λ 531, *i.e.*, near the corona line. In Vogel's observations, the line is always associated with the iron lines E and 519, and as great accuracy is not claimed for the measures, it is quite possible that it is simply the third brightest flame line of iron at 5327. In the map it is represented as being fainter than 527, and this is quite consistent with this origin. Cornu represents the line without E or 579, and suggests that it is identical with the corona line.

The line near 635, recorded by three observers, is no doubt the same as that seen in the spectra of the bright line stars of Cygnus. No terrestrial equivalent has yet been found, but a line in this position is seen in the spectrum of the Limerick meteorite.

It is thus seen that practically all the lines and bands in Nova Cygni can be explained by reference to laboratory work.

It is extremely interesting to attempt to build up the spectrum on any given date by integrating the spectra of comets and nebulæ, or nebulæ and bright-line stars, which we have reason to believe to be swarms of meteorites. Some of these integrations are shown in the diagrams (Figs. 59, 60, 61). As in Novæ we have to deal with collisions of swarms of meteorites of different densities, in such integrations it is necessary to take swarms of different degrees of condensation.

The first integration shown (Fig. 59) is for the spectrum of December 8, as observed by Vogel. In this case the spectrum is reproduced almost line for line by adding together the spectra of a nebula, Coggia's Comet (June 13, 1874), and the Great Comet of 1882 at perihelion (brightest lines only). The only

[1] *Proc. Roy. Soc.*, vol. xlviii, p. 200.

lines of importance which are in the Nova and not in the integration are C and a line near 635, and we know that these are associated with the spectra of meteor-swarms; C occurs in γ Cassiopeiæ and β Lyræ, and the line at 635 occurs in the bright-line stars in Cygnus. This integration strengthens the view that Nova Cygni was produced by the collision of swarms of different densities. In such a case, there would first be the collisions between the two sets of outliers, then the denser part of the smaller swarm would enter the outliers of the larger, and finally the densest parts of both swarms would come together; so that, in this way the resulting temperature effects may be very complicated.

The spectrum of the Nova on a later date, January 27, 1877, is considerably simpler than that of December 8, and may be almost perfectly reproduced by integrating the spectrum of a planetary nebula and that of γ Argûs (Fig. 69). All the lines in the Nova on this date are shown in the integration, with the exception of the C line. The integration, however, in this case shows the lines at 495 and 568, which were not recorded in the Nova; but their absence was probably only due to difficulties of observation. The line at 495 is excessively faint in some nebulæ.

The spectrum of the Nova on March 2 was still simpler than on former days, and can be fairly reproduced by adding together the spectrum of the bright-line star, Lalande 13,412, and that of a planetary nebula which shows only the chief line of the nebulæ. This is shown in Fig. 70. The temperature in this case is obviously very much lower. We get only the lowest temperature line of iron at 579, of magnesium 500, the F line of hydrogen, and a slight indication of carbon. The absence from the Nova of the line near 540 is probably also due to difficulties of observation.

It will thus be seen that the complete discussion of Nova Cygni generally bears out the statement I ventured to make

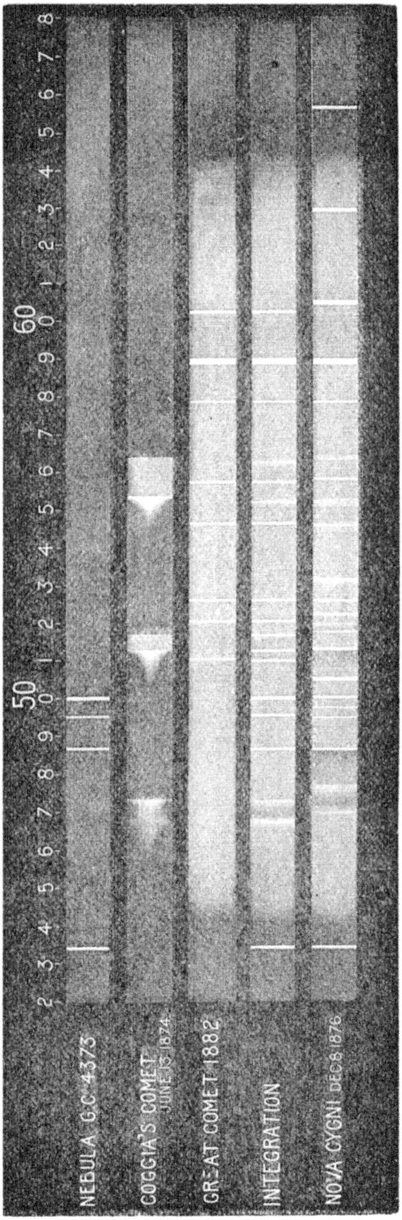

FIG. 59.—Integration of the spectra of comets and nebula, compared with that of Nova Cygni.

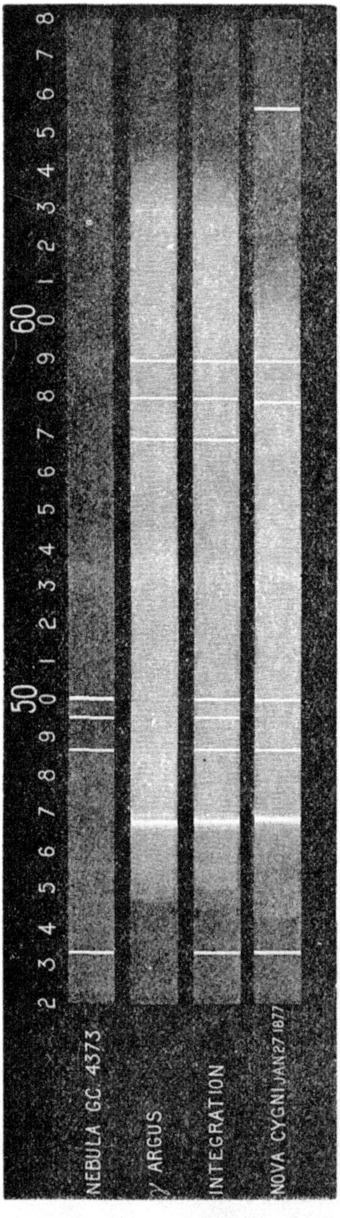

Fig. 60.—Integration of the spectra of nebula 4373 and γ Argûs, compared with that of Nova Cygni.

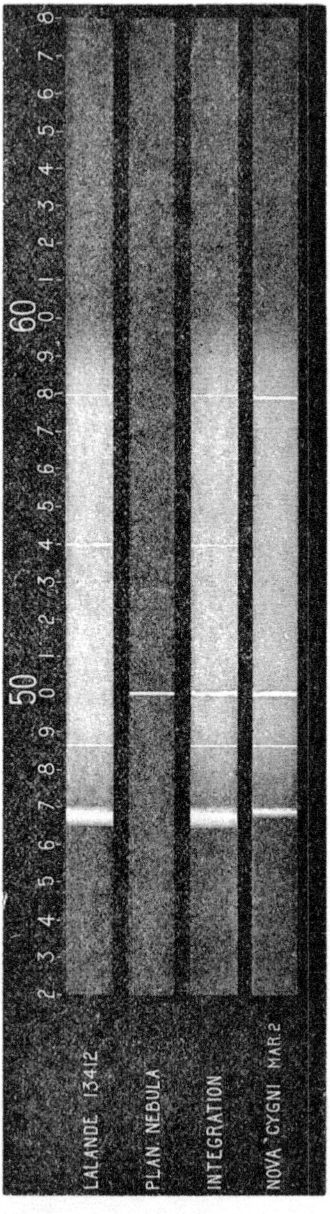

Fig. 61.—Integration of the spectra of a planetary nebula and Lalande 13,412, compared with that of Nova Cygni.

after the preliminary inquiry in November, 1887, viz.: "This star passed through all the changes of temperature represented by stars with bright lines, comets, and nebulæ."

The magnitude of this star varied from 3 on November 24, 1876, to 11½ on February 18, 1878.[1]

Nova Andromedæ, 1885.

The spectrum of Nova Andromedæ was first examined by Dr. Copeland on September 1, 1885, when it was found to be[2] "continuous from end to end, and only on close examination could slight condensations, indicative of bright lines, be detected. The spectrum was not considered to differ strikingly from that of the nebula." At this time the star was "yellowish" in colour, on September 3 it was "full yellow." A close spectroscopic examination, however, was not possible until September 10, when Dr. Copeland wrote :—

"With the unmagnified dispersion of a direct-vision Vogel spectroscope the spectrum extended from W.L. 670 mmm. to 453 mmm., or, from between B and C to half-way between F and G. When the spectrum was sufficiently narrow all the colours were visible, with a suspicion of brighter points in the line. An attempt was made to measure these with the Grubb spectroscope and a flint prism of only 40° refracting angle. The instrumental change cut down the spectrum to the limits of 600 mmm. and 456 mmm., with a maximum at 544·41 mmm., and a suspicion of a bright line, but hardly more at 482·2 mmm. With the same apparatus the spectrum appeared *quite continuous* on September 11, but again showed traces of bands on the 13th, and was slightly banded on the 15th. Traces of a condensation of light were seen at W.L. 471·6 mmm. on September 20."

Using a special acute prism of only 15° angle, it was noted: "Only traces of two brighter points towards the yellow end of the spectrum could be made out on September 30, the rest of the spectrum appearing absolutely continuous." On October 1,

---

1 *Ast. Nach.*, vol. lxxxix, 1877, p. 42. *Berlin Akad. Monatsber.*, 1877.

2 *Monthly Notices, R. A. S.*, vol. xlvii, p. 50.

these two lines and another fainter one were measured, the mean wave-lengths being as follows, 546·8, 514·0, 489·2.

Dr. Copeland thought that some of the discordances of the measures were due to the indeterminateness of the objects measured, and noted that, making due allowance for this uncertainty, it seems probable that the three "bright" bands are identical with the three brightest bands afterwards measured with the same apparatus in Mr. Gore's *Nova Orionis*, of which the brightest parts were at wave-lengths 542·8, 516·2, and 494·4.[1] The trace of a condensation of light at W.L. 471·6, seen on September 20, agrees well with the bright line in Nova Orionis at W.L. 472·2; while the maximum of light in Nova Andromedæ at 544·4, on September 10, is closely in accord with that for the star in Orion at 542·8. The only really discordant item is the point laid down at 482·2 on September 10, which does not correspond to any known bright band in the spectra of variable stars; it was, however, entered in the note book . . . as "a suspicion of a bright line, but hardly more." But if it does roughly represent the position of a "bright" line actually visible in the spectroscope, one would feel inclined to regard it as a trace of the F line. On October 2, the spectrum presented the same appearance as on the preceding day. On October 19, it could be still noted as continuous, but not uniform, with the Vogel spectroscope. Dr. Copeland remarks (p. 55):—

"Although the foregoing results differ widely from those obtained at Greenwich and also at Yale College" [which will be given subsequently], "as regards the three chief lines the observer cannot doubt as to their general correctness. . . . . Besides, in the case of an object losing so rapidly its power of emitting light, it is quite possible that the spectrum may have slightly varied from day to day

"In conclusion, it seems worthy of remark that the spectrum described above is the same as that given by any ordinary hydrocarbon flame, burning so feebly that the spectrum of the blue base of the flame is just beginning to show through the continuous spectrum afforded by the white part of the flame."

---

[1] *Monthly Notices, R. A. S.*, vol. xlvi, p. 110.

On September 1, Professor Vogel observed the spectrum and noted that the red and yellow were bright and the green faint. He also found dark lines or bands, one between the yellow and green, and another between F and G.[1]

Dr. Huggins communicated his observations of Nova Andromedæ to the *Observatory* for October, 1885. They are as follows:—

"The star was observed here first on the night of the 3rd instant. It presented the appearance of an orange-coloured star of from the 8th to the 9th magnitude. With a spectroscope of low dispersive power, a continuous spectrum was seen from about C in the red to a little beyond F. There was an apparent condensation of light from about D to *b*, which might be due to bright lines in that part of the spectrum. This supposition was strengthened by the employment of a more powerful spectroscope, but I was not able to be certain on this point.

"On the night of the 9th instant, the star, which was then distinctly on one side of the principal point of condensation in the nebula, appeared to me to have a less decided orange tint. It presented a similar appearance in the spectroscope, with the exception that the light was less strong about D. I was, so far, confirmed in my suspicion of bright lines, that I have little doubt that from three to five bright lines were present between D and *b*."

On September 4, Mr. Maunder noted that—[2]

"The star gave a perfectly continuous spectrum, in which no lines, either bright or dark, could be detected. The red, orange, and violet were very faint, or altogether wanting, the spectrum being traceable from about D

---

[1] "Die Beobachtungen am 1. and 2. September über den Stern in Nebel ergeben, dass derselbe auch bei stärkerer Vergrösserung (550 f.) vollkommen sternartig bleibt, und dass das Spectrum continuirlich ist. Die Intensitäten der Farben im Spectrum scheinen etwas abweichend von den gewöhnlichen Sternspectren zu sein, indem Roth und Gelb besonders stark hervortreten, Grün aber verhältnissmässig schwach ist. An der Grenze des Gelb und Grün habe ich eine dunkle verwaschene Bande vermuthet, eine zweite ebensolche im Blau zwischen F und G. Im Fall der Stern noch heller werden sollte, denke ich Sicherheit über weiteres Detail, welches ich im Spectrum vermuthete, zu erlangen. Ich bemerke noch dass der Andromeda-Nebel continuirliches Spectrum giebt und dass der Kern des Nebels h. 51 ein Spectrum zeigt, was mit dem des neuen entstandenen Sterns in h. 50 übereinzustimmen scheint." (*Ast. Nach.*, No. 2681.)

[2] *Monthly Notices*, *R. A. S.*, vol. xlvi, p. 19.

to F, but being scarcely discernible beyond those limits. The spectrum of the star resembled that of the nebula precisely, except, of course, that it was brighter, and probably in consequence of this greater brightness, was traceable a little further in both directions.[1]

Konkoly, writing on September 5, noted that he observed the spectrum of Nova Andromedæ on the previous day.[2] In his words:—

"Yesterday, when I examined the star, it seemed of a reddish-yellow colour and very faint. I estimated its diameter at, at least, two seconds; it appeared as a nebulous star. Its spectrum was unexpectedly faint, so that it was impossible to use a cylindrical lens. The edges of the spectrum seemed to be enveloped in a coloured mist. It gives the impression of bright fields on a dark ground, and in the red, yellow, green, and blue a broad band is seen. If this be the case, these broad bright regions would correspond to the hydrogen lines C, F, as well as D^3, and at an enormously high pressure. A similar broad field is also seen in the green, which certainly cannot belong to the group named. I should be inclined to class the spectrum with Type III*b*, and Professor Ritter coincides in this view.

"It must not be overlooked that the violet part is absent in the spectrum; and shortly behind the region of F the spectrum is just as if cut off. This observation was confirmed by Dr. v. Kovesligethy."

Mr. O. T. Sherman, of Yale College Observatory, also made some observations.[3]

Some observations of the Nova were made by Dr. Lohse with the 15·5-inch Cooke refractor, at Mr. Wigglesworth's Observatory.[4]

On September 3 it was noted,—

"The spectrum is continuous, no lines could be distinguished in it with the Browning and Maclean spectroscopes. The bright C line was very well seen in γ Cassiopeiæ."

The late Rev. S. J. Perry recorded:—[5]

"The Nova in Andromeda has been observed at Stonyhurst on every favourable occasion from September 13 to November 8. On September

---

[1] For observations in detail, see my paper p. 427.

[2] *Ast. Nach.*, No. 2681.

[3] *Ibid.*, No. 2691.

[4] *Monthly Notices, R. A. S.*, vol. xlvi, p. 299.

[5] *Ibid.*, p. 22.

13, the spectrum was found to be continuous, but the red end was absent, and there was a decided maximum in the green. A bright band in the green was suspected, but not clearly seen. The spectroscope was carefully focussed on the lines in the spectrum of β Andromedæ, which were well seen, before being turned on the nebula of Andromeda. On October 9 the spectrum was still brightest in the green."

The following table brings some of these observations together[1] for reference:—

Date.	Observer.						
1885.							
Sept. 10	Copeland[1] .	—	482	—	—	544	—
,, 11	,, [2] .	—	—	—	—	—	—
,, 11	Sherman [3] .	—	486	—	527–536	—	551–564
,, 11	Maunder ..	—	—	—	5327	548	—
,, 20	Copeland..	471·6	—	—	—	—	—
,, 30	,, ..	—	—	517	—	546	—
Oct. 1	,, ..	—	489	517	—	546	—
,, 2	,, ..	—	489	517	—	546	—
Suggested origins......		C 468–474	H 486 (F)	C 517	Fe 527	Pb 546	Mn 558

[1] Light less strong about D.
[2] "Quite continuous."
[3] Spectrum sharply terminated about D.

From the foregoing it will be evident that the observations were extremely difficult throughout, owing partly to the dimness of the star, and partly to the character of the spectrum. It must also be pointed out that the spectrum observed was the integrated effect of the spectrum of the Nova and the spectrum of the nebula itself, the nature of which I pointed out in 1888.[2] Even the observation of the spectrum of the nebula is one of considerable difficulty, especially when the sky is at all hazy. It then appears simply as a faint continuous band of light, but when the sky is clear it is seen to have at least three maxima. It is evident from the observations of the Nova, that on some occasions the bands in the spectrum of the nebula were observed

[1] A complete list of all observations is given in my complete paper, p. 430.
[2] *Proc. Roy. Soc.*, vol. xlv, p. 215.

in addition to those special to the Nova, whilst on others they were not. This appears to have been especially the case in the later observations when the Nova had become dim, and this is, of course, what would be expected.

I wrote in 1888:—[1]

"We have seen that some planetary nebulæ give the same spectrum as a comet at aphelion. It appeared that if the Nebula of Andromeda were further advanced than a planetary nebula in condensation, it should give a spectrum approximating to one of the more advanced cometary stages which have been already discussed.

"The spectrum of this nebula has hitherto been regarded as a perfectly continuous one, but the observations referred to show that there are some parts brighter than others. The spectrum is almost entirely wanting in red and yellow light.

"In the green there are two maxima, the brightest of which is at wave-length 517, as near as could be determined with the wide slit which it was necessary to employ, the other maximum is near 546. One of the observers, Mr. Fowler, made six independent measures of the maxima on November 20, and got very nearly the same result each time, comparison being made with the spectrum of a Bunsen, and the spectrum of chloride of lead at the temperature of the Bunsen. The measurements were repeated on November 27 with the same result, and on this occasion they were confirmed by another observer, Mr. Coppen. Another brightness near 474, as determined by comparison with the Bunsen burner, was also suspected, but it was not so easy to measure as the others.

"My suggestion as to the origin of this spectrum is that it is the integration of very slight continuous spectrum, carbon fluting radiation, and the absorption of manganese (558) and lead (546). The citron band of carbon masks, and is masked, by the manganese fluting, and the absorption fluting of lead causes, by contrast, the apparent brightness at 546. The brightest maximum is no doubt the brightest fluting of carbon at 517, and the one in the blue, which was suspected, is probably the blue carbon group 468–474.

"If these observations are confirmed, this nebula is at present at the same stage of condensation as Comet I, 1868, on April 29 (P.P., April 20), which must be regarded as a pretty advanced cometary stage, seeing that it was observed so near perihelion and that the perihelion distance was small."

"The discussion of the observations of Nova Andromedæ, which is not yet completed, shows that there were bright lines in exactly the same

---

[1] *Proc. Roy. Soc.*, vol. xlv, p. 215.

positions as the brightnesses which have now been determined in the nucleus of the nebula. The appearance of the Nova was, therefore, probably due to increased temperature, due to collisions taking place between the sparser outliers of the swarm composing the nebula and an external swarm which came in contact with them. The view of the Nova's connection with the nebula is, therefore, greatly strengthened by this inquiry."

These observations have since been confirmed by Mr. Taylor,[1] who observed two brightnesses at 517·4 and 547·3, and suspected one in the blue.

The only lines seen in the Nova, which are not seen in the nebula, are F, 531 and 558, with, perhaps, D^3 and C. The spectrum of the nebula shows manganese 558, and lead 546 absorption, masking the carbon at 564, and making it appear at 546. The Nova spectrum added manganese radiation 558. The line near 530 I have taken as E (527), the lower temperature line of iron at 579 not being visible on account of the greater brightness of the continuous spectrum in that region. The hydrogen lines C and F were probably due to collisions of the outliers of the two swarms. Mr. Sherman[2] recorded F as a line in the nebula, whilst Mr. Maunder noted that the line at 548 could be traced over or near the nucleus of the nebula, thus, to a certain extent, confirming the Kensington observations. There is also evidence in the observations to show that the continuous spectrum fluctuated in relative brightness during the visibility of the Nova. The effect of a brightening of the continuous spectrum would be to mask the faint lines in the green or yellow whilst affecting but little any that might occur in the blue. Lohse's observation of a "faint part, midway between F and G," in the absence of other lines, is a case in point. It is not unlikely that this was really the blue carbon band which is seen in the nebula itself, although its position was only roughly estimated.

[1] *Monthly Notices, R. A. S.*, vol. xlix, p. 126.
[2] *Ast. Nach.*, No. 2691.

Again, on September 20, Dr. Copeland observed traces of a condensation of light at wave-length 471·6, and this also was probably the blue carbon band of the nebula, other lines being masked by the brighter continuous spectrum. On other occasions, not even the faint blue band was seen, the spectrum being recorded as quite continuous.

As the brighter continuous spectrum due to the Nova gradually dimmed, the bands of the nebula in the green became more prominent.

It is thus seen that although there does not appear to have been a regular sequence of events in the spectrum of Nova Andromedæ, as far as the actual observations go, it is probably due to difficulties of observation, and to the fact that the spectrum of the Nova was superposed upon that of the nebula.

The apparent variations from day to day are possibly not all real, and it is hopeless to attempt to explain them all by reference to the effects of a gradual fall of temperature. The star only diminished about two magnitudes during the period in which spectroscopic observations were made, and hence the change of temperature would not be so great as in Nova Cygni, and consequently the variations of spectrum would not be so evident.

No lines or bands were recorded in the spectrum with which we are not familiar in other bodies.

The line at λ 546, which was seen in the spectrum of the Nova, was undoubtedly the maximum of light, probably due to lead, which is seen in the nebula itself. This is also the probable explanation of the other lines near λ 517 and λ 473, the latter being only observed on one occasion. The additional lines due to the collisions, which produced the Nova, were, therefore, F, 5327 and 558, if we neglect Konkoly's doubtful observation of C and D^3. The appearance of the hydrogen line F is exactly what would be expected from what we know of its appearance in such stars as Mira, when, by additional collisions

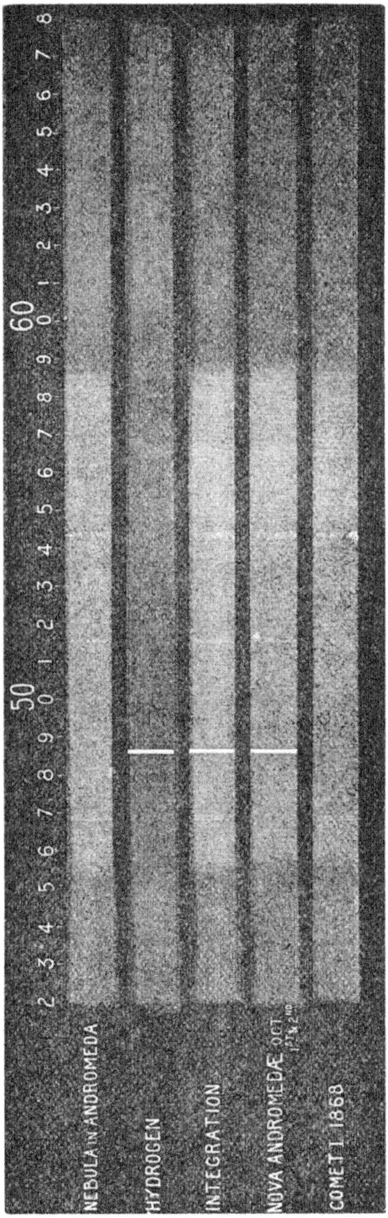

FIG. 62.—Integration of spectra of the nebula in Andromeda, and of hydrogen, compared with that of Nova Andromedæ.

due to the periastron passage of a revolving swarm, the star reaches a maximum. The line near λ 558, seen by Maunder and Sherman, was in all probability due to the brightest manganese fluting at the same wave-length, which is very frequently recorded in cometary and other spectra. This, and the line 5327, which was most likely the iron line E (λ 5268) were probably produced by local collisions in the denser parts of the swarm.

Fig. 62 will show how the spectrum of the Nova, as seen by Copeland, on October 1 and October 2, can be reproduced by adding the spectrum of hydrogen to that of the Andromeda nebula. The third band of the nebula was not seen, but this is always most difficult to observe. At this time, then, the only line special to the Nova was F, and this, it will be remembered, was seen last in Nova Coronæ.

The star when first seen on August 19, 1885, at Belfast by Mr. Ward, was of the ninth magnitude; it decreased to thirteenth magnitude by February, 1886.

The Sequence of Phenomena in the Spectra of Novæ.

Sequence of Spectra.

If the appearance of a new star be due to the collision of two meteor swarms as I have suggested, it is obvious that the spectroscopic changes should follow the same order as those observed in the spectrum of a comet during its passage from perihelion to aphelion, when differences of observing conditions, and the relative physical conditions of the two swarms which produce a Nova, are duly allowed for.

The following map (Fig. 63) shows the theoretical sequence from this point of view, commencing in the first horizon with one swarm sufficiently sparse to give only the bright lines of hydrogen and flutings of carbon, and the other sufficiently dense to give dark D and *b* and the flutings of lead, manganese,

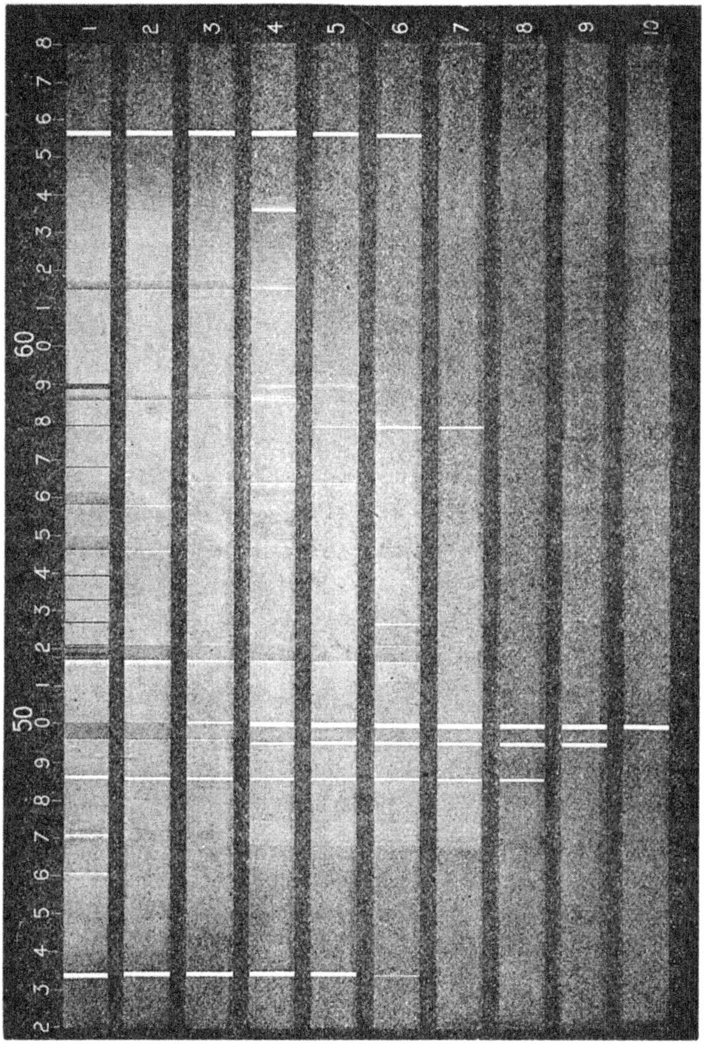

Fig. 63.—Sequence of spectra in Novæ (relative intensities).

and iron, with bright carbon flutings. The first horizon accordingly shows the integration of these two spectra, and subsequent ones the effects of cooling of the mixed swarm.

None of the four Novæ, which have been spectroscopically

examined, have shown the complete sequence of changes indicated in the map, but Nova Cygni passed through most of them.

As the Nova decreases in temperature, and, therefore, in light, the line absorption disappears and the metallic fluting absorption decreases in intensity. Carbon radiation, and the hydrogen lines, C, F, and G, remain about the same as before horizon (2).

The next stage (3) brings us to a condition represented by a swarm of Species 3, Group II.[1] The manganese and lead absorption at 558 and 546 respectively, are now overpowered by the carbon radiation at 564, whilst the absorption of the second manganese fluting at 586 is still apparent. The hydrogen lines, C, F, and G, remain almost as before.

On the following horizon (4) is represented the radiation of lines and flutings. The manganese and lead radiations are visible. The brightest iron and magnesium lines are seen; also sodium, D, and the line at 495, whilst the brightest edge of the magnesium fluting at 500 is just visible. This was a condition observed by Vogel in Nova Cygni, on December 8, 1876. It was also observed in the Great Comet of 1882, when near perihelion.

The condition following this (5) is that in which lead and manganese radiation have disappeared; the three carbon flutings are well seen, C, F, and G are rather fainter, but the fluting at 500 appears brighter. The 495 line and the iron line 579 are still seen, the other iron lines in the green being masked by the brightness of the continuous spectrum. This condition was approached in Nova Cygni on February 2.

On the following horizon (6) the 564 carbon fluting is not visible, and the 517 fluting appears almost as a line. The iron lines, 579 and E, are now visible in consequence of the fading

---

[1] *Bakerian Lecture*, 1888, p. 66.

of the continuous spectrum. The hydrogen lines are still visible, whilst D has disappeared. The two magnesium flutings, 5210 and 500, are each represented, the latter being the brighter. This stage in the sequence was observed in Nova Cygni on February 2, and the brightest of the lines in Nova Coronæ, May 19. All the lines have been recorded in the spectrum of the nebula in Orion.

Both the carbon flutings, 517 and 564, now disappear (7), having become so pale that they are masked by the continuous spectrum, whilst the 468 maximum of the carbon fluting is visible, because the continuous spectrum does not reach it. F is the only remaining hydrogen line, and 579 the remaining iron line, this being the line visible at the lowest temperature. 495 is rather fainter, whilst 500 has increased in brightness. This stage was observed in Nova Cygni on March 2, 1877. The 474 carbon is the next to disappear, and three lines only are left:— F, 495, and 500 (8). The nebula G.C. 4373 and many others give this spectrum. No observations of Nova Cygni were made between March and September, 1877, and it is between these dates that this stage would have occurred.

With regard to horizon (9), the nebula G.C. 2343 and many others give a spectrum consisting of two lines, 495 and 500, and Lord Lindsay observed in Nova Cygni, on October 10, 1877, a line at 492 (as well as the line at 500), which was most probably the nebula line at 495.

The last stage in the sequence is when the 500 fluting remains alone (10). This was observed in Nova Cygni by Vogel and by Copeland; it is the only line in the G.C. 4403, and, as I have shown in the appendix to the Bakerian Lecture, it is the characteristic line of comets when at a great distance from the Sun, and was observed by Dr. Huggins in 1866 and 1867.

It should be remarked that this is only one hypothetical case of a Nova, and that there may be considerable variations from it, according to the initial conditions of the two swarms which produce the Nova by collisions. Thus, in the case of Nova Andromedæ, one of the swarms already existed as a nebula and the collision with the other swarm only sufficed to bring out a few more lines, including the F line of hydrogen. Here, then, the initial spectrum would be different, and the subsequent changes would not take place in exactly the same order. The integrations which have been given, however, show that it is possible in any of the Novæ which have yet been spectroscopically observed to get a good idea of the states of disturbance of the two swarms after the first collision.

It is important to compare this sequence with that which I have already given for comets,[1] for, though the conditions are different, there will be a certain similarity, since both have to be regarded as meteor swarms. On the first three horizons of the Nova sequence we have mixed radiation and absorption phenomena; this also occurs in comets, though the difference in the brightness of the continuous spectrum does not make it so obvious. On the fourth horizon of the Nova sequence we have carbon, manganese, and lead radiations, which also occur in comets; but, in addition, there are lines of hydrogen and other substances due to the compound character of the swarm. Finally, both Novæ and comets give one line, the chief line of the nebulæ, which is probably due to magnesium. It will be seen that the spectra of Novæ are, in general, more complex than those of comets, which result from the collision of two swarms of different densities.

Variations in Magnitudes.

In each case where the spectra of new stars have been observed the evidence tends to show that the star was hottest at the first observation; the absorption lines giving way to

---

[1] *Proc. Roy. Soc.*, vol. xlv, p. 190; *Meteoritic Hypothesis*, p. 211.

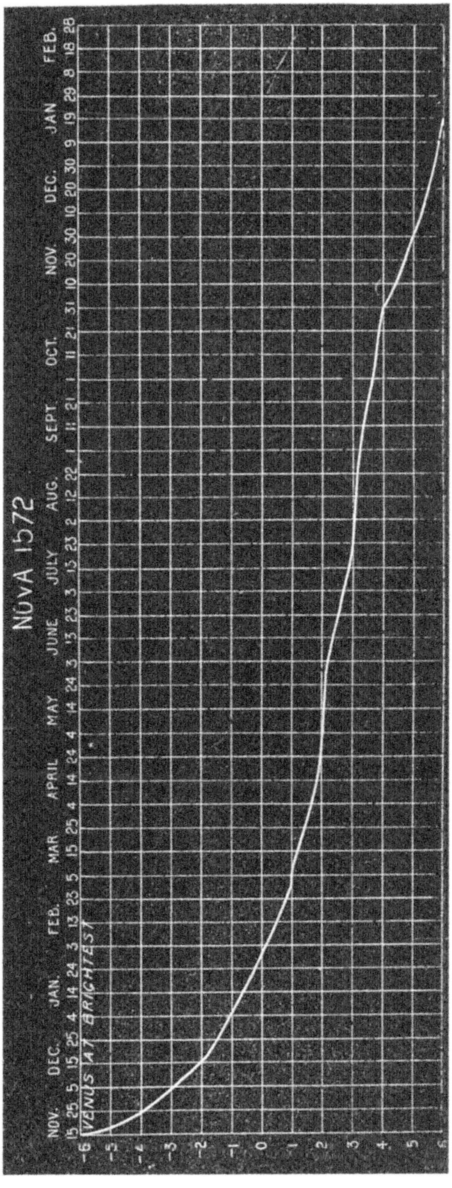

FIG. 64.—The light curve of the Nova of 1572.

bright lines in Nova Coronæ, the brightest lines fading away one by one in Nova Cygni, and the carbon becoming more manifest in Nova Andromedæ, all go to show a diminution in the temperature of the star after the first observation. Indeed, in only one case, that of Nova Andromedæ, have I been able to find any evidence of a new star increasing in brightness after its discovery. Assuming that the Nova was physically connected with the nebula, this increase of brightness is exactly what would be expected from a consideration of the beautiful photograph obtained by Mr. Roberts, which shows that the nebula is really a spiral system. We have only to suppose the incoming swarm to pass from the outside to the inside of one of the spirals—a region of gradually increasing density—to give the required explanation.

It is very probable that a Nova would be overlooked until transcendentally bright, and the observations of the magnitudes of Novæ show that such has been the case. It is, however, essential to my theory that the increase in temperature and in luminosity shall be much more sudden than the decrease.

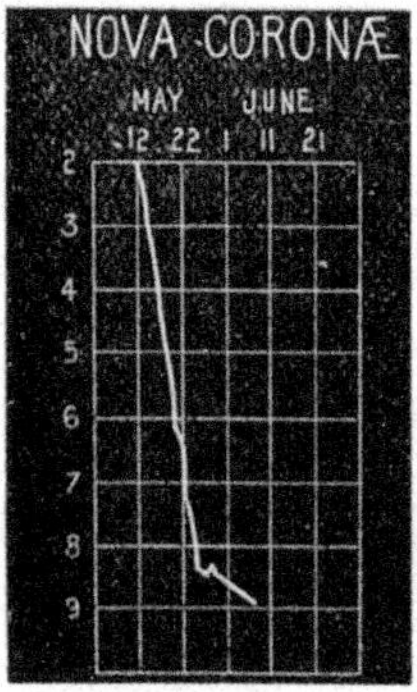

FIG. 65.—The light curve of Nova Coronæ.

The first observations of Nova Coronæ, which showed the same absorption lines as α Orionis, indicate a comparatively high temperature, and it also was a Nova that flashed out very

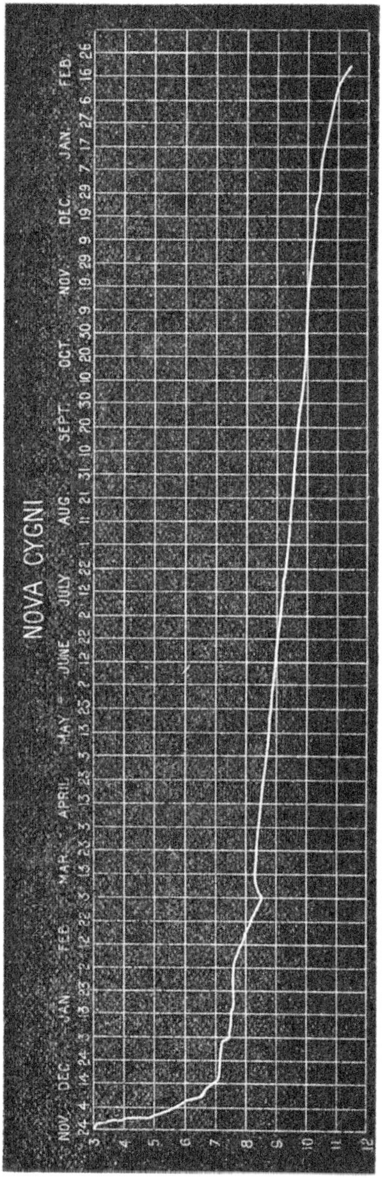

FIG. 66.—The light curve of Nova Cygni.

suddenly. M. Courbebaisse and Mr. Baxandall[1] held that the Nova could not have been conspicuous two days before its discovery, and were confident it was not visible three days before. Many other observers support this statement. Hence it is evident that in these cases we are dealing with the collisions of two rather condensed swarms of meteorites, the consequent temperature being high and the increase in light sudden. In Nova Andromedæ, where the increase in luminosity was not so sudden, the temperature was not nearly so high. In this case we had most probably to deal with the collision of two swarms not nearly so disturbed as in Nova Coronæ; perhaps a slightly condensed swarm passing through the Andromeda nebula, in which case the increase in temperature would be more gradual and comparatively small.

Nova Cygni decreased in magnitude in a very similar manner to Tycho Brahe's Nova, dimming very suddenly at first, and more slowly later on. In three months the Nova fell from the 3rd to the 8th magnitude, and then the fall from 8 to 11 took twelve months. The brightness of this Nova, as well as its long period of visibility afforded the opportunity for many observations of its spectrum, the result being that the sequence of phenomena, as far as it goes, is much more complete in it than in any other Nova, which brings us to the conclusion that other Novæ would have given a more complete sequence if more spectroscopic observations could have been made.

---

[1] *Monthly Notices, R. A. S.*, vol. xxvi, p. 293.

CHAPTER XIII.—NOVA AURIGÆ.

THE FIRST GLIMPSES.

As I said at the commencement of the last chapter, the discussion there given was based upon a paper communicated to the Royal Society in 1891.

Fortunately for science, another new star appeared in 1892; it was known as Nova Aurigæ, and two photographs will give us an idea of the sort of thing which an astronomer sees in the heavens when the discovery of a new star is announced. The photographs show a portion of the constellation of Auriga, and a star which is very clearly seen in the photograph taken very soon after this star had burst upon us, is absent from one taken a few months later.

1

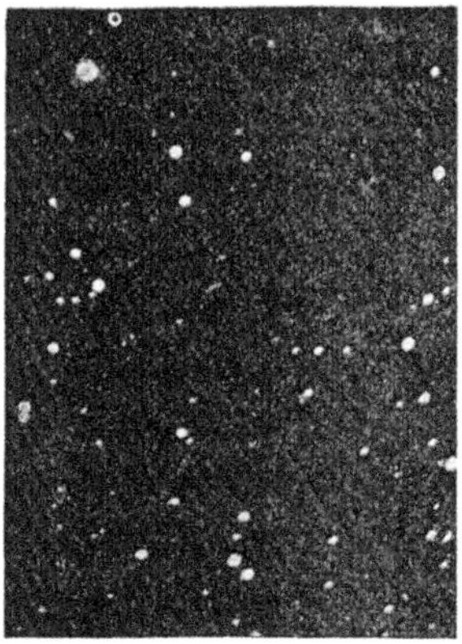

2

FIG. 67.—The region in the heavens where Nova Aurigæ was observed (1) after its disappearance; (2) when brightly visible (nearly in the centre).

It will have been gathered from the previous chapter that since the spectroscope was first applied to the stars, three new stars had been observed and spectroscopically examined. One appeared in Corona Borealis in 1866, one in Cygnus in 1876, and one in Andromeda in 1885 ; then came one in Auriga in 1892, and last of all was one in the southern hemisphere, discovered in 1893. The first three of these were observed by eye only, but in the two recent ones we had the immense benefit of photographic records.

It was therefore a very interesting point when a new star came along, that we could examine with the powerful instruments used in modern research, to see whether there was any additional light thrown by it upon the problem of two bodies, and especially upon one point, in which, if the meteoritic hypothesis failed, it was worth absolutely nothing at all. If there was any truth in the idea of the light of these bodies being produced by the clash of meteor-swarms, when the clash was over the swarms should go back into their native obscurity, or condition of low temperature, and should, if they were seen at all, put on the spectrum of sparse swarms in other parts of the sky; that is, they should put on the spectrum of a nebula.

The appearance of Nova Aurigæ furnished, indeed, a splendid opportunity of testing the many theories which have been at various times advanced to account for the phenomena. This Nova was discovered at Edinburgh by Dr. Anderson, who was modest enough to announce his discovery by sending an anonymous post-card to Dr. Copeland, the Astronomer Royal for Scotland, on February 1, 1892. It was then a star of the 5th magnitude, and on confirming the true nature of the newly discovered star by means of the spectroscope, Dr. Copeland made the news public.

During the next two or three weeks the star fluctuated considerably in brightness, though being generally on the down grade; and by April 26 had fallen to the 16th magnitude, so that it could only be picked up in the very largest telescopes. Thanks to the photographic records of stellar spectra taken at Harvard, it was possible to learn something of the earlier history of the new star. It had really been photographed by Professor Pickering two months before its existence was known.

Up to December 1, 1891, the light of the Nova did not equal that of a 5th magnitude star on the photographic plates, but from December 10 to January 20, 1892, the star was photographed.

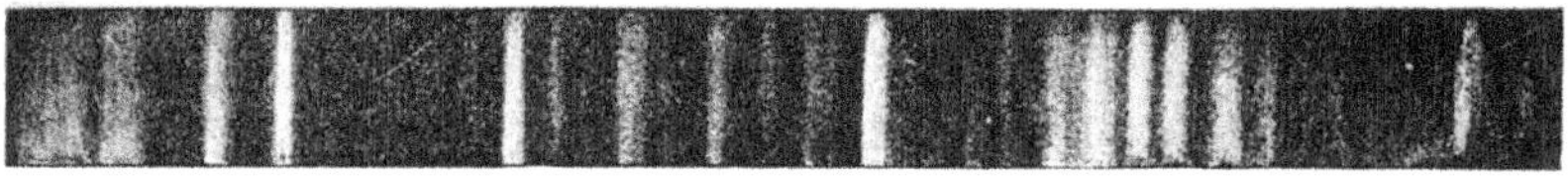

FIG. 68.—Photograph of the spectrum of Nova Aurigæ, taken at South Kensington, February 7, 1892.

Fig. 68 reproduces a photograph of the spectrum of this wonderful star, taken at Kensington.

I learnt from a note in *The Times* of Wednesday, February 3, that a new star had been discovered. The night was fortunately fine, and two photographs of the spectrum were taken with the same instrument which had been employed to obtain stellar spectra, a 6-inch refractor with a large prism in front of the lens. The first photograph contained thirteen lines, the second more; the exposures were necessarily long.

I shall refer to the wave-lengths of the lines later on.

On the photographs it was noticed that several of the bright lines were accompanied on their more refrangible sides by dark lines, but as the matter was so important, no announcement was made till further confirmation had been obtained.

Now, the same set of particles cannot be producing bright and dark lines at the same time. We were then obviously dealing with two sets, and the photographs, therefore, which were taken of the spectrum of this strange body, put beyond all question the fact that we were really dealing with two bodies, and not with one. That was very important; we see from the photograph that the spectrum of the Nova first obtained was very unlike the spectrum of nebulæ, so that it required a certain amount of faith, when the spectrum was observed, to suppose that after a certain time, when the action which produced the greater luminosity was reduced and the light toned down, we should eventually get the spectrum of a nebula.

But this is to anticipate. Let us first deal with the early photographs and observations. With regard to these, I must chiefly rely on the series made at Kensington, for I have not yet had sufficient leisure to bring all those made by others together.

For the eye observations, the new 3-foot mirror, figured by Dr. Common, was employed, but unfortunately the clock was not mounted, so that the observations were difficult.

C was the brightest line observed. In the green there were several lines, the brightest of which was in all probability F, the position being estimated by comparison with the flame of a wax taper. Another line was coincident, with the dispersion employed, with the radiation at λ 500 from burning magnesium wire. A fainter line between the two last named was probably near λ 495, thus completing the trio of lines which is characteristic of the spectra of nebulæ. There was also a fairly bright line or band coincident with the edge of the carbon fluting near λ 517 given by the flame of the taper. A feeble line in the yellow was coincident, under the conditions employed, with the sodium line at D.

The colour was estimated by Mr. Fowler as reddish-yellow, and by Mr. Baxandall as rather purplish. My own impression was that the star was reddish, with a purple tinge ; this was in the 10-inch achromatic. In the 3-foot reflector it was certainly less red than many stars of Group II. No nebulosity was observed either in the 3-foot reflector or the 10-inch refractor; nor did any appear in a photograph of the region taken by a 3½-inch Dallmeyer lens with three hours' exposure.

On February 7 two more photographs were taken; though exposed for a shorter time than the previous ones, they gave many more lines, and supplied ample confirmation of the fact that the bright lines at K, H, *h*, and G were accompanied by dark lines on their more refrangible sides.

This important discovery was communicated to the Royal Society the next day (February 8). I may add that I learnt from *The Standard* newspaper of February 10 that the same appearance had been observed at Harvard College Observatory.

Dr. Vogel saw the doubling on February 14.[1]

THE PHOTOGRAPHS.

On the 7th the first photograph was exposed for 1½ hours and the second for two hours. The same number of lines is shown in both photographs. Twenty bright lines were measured by Mr. Baxandall, and their wave-lengths determined on Rowland's scale. I shall refer to these in the sequel.

In addition to the lines recorded in the table, the photographs of the spectrum of the Nova showed several lines more refrangible than K. They probably include some of the ultra-violet hydrogen lines.

Many of the lines in the spectrum of the Nova were broad although in a photograph of the spectrum of Arcturus, taken with the same instrumental conditions, the lines were perfectly sharp. The broadening of the lines was not accompanied by any falling off of intensity at the edges, as in the case of the hydrogen lines in such a star as Sirius. With the method employed in taking the photographs, long exposures are liable to result in a thickening of all the lines, on account of atmospheric tremors. The lines would also be thick if the

---

[1] *Ast. and Ast. Phys.*, 1894, 897.

Nova be hazy, as observed at Greenwich. In the photographs, however, all the lines are not equally thick. I pointed out on February 11 that, if the lines are similarly broadened when a slit spectroscope is employed, the effect must be due to internal agitations; for if different regions of the Nova were moving with varying velocity, or with the same velocity in different directions, a normally fine line might be widened, as observed in the photographs.

The hydrogen lines and the K line of calcium were very bright.

Photographs of the spectrum were attempted on several subsequent dates. Those of February 11, 12, 16, and 23, however, were insufficiently exposed; still they showed that the dark lines were still more refrangible than the accompanying bright ones, and that the same lines were present as in the previous photographs. A plate was exposed for 2 hours 35 minutes on February 24, but no impression was obtained. The photograph taken on February 13 was identical with those already referred to. In the three photographs of February 22 there appeared to be a slight diminution in the intensity of the H and K lines, but otherwise there was no decided change.

The Eye Observations.

On February 7, with a 10-inch refractor and Maclean spectroscope, C was seen to be very brilliant, and there were four very conspicuous lines in the green. Several fainter lines were also seen, and a dark line was suspected in the orange. I noticed that some of the lines, especially the bright one near F, on the less refrangible side, appeared to change rapidly in relative brightness, and this was confirmed by Mr. Fowler.

Observations of the spectrum were made by Mr. Fowler with the 3-foot reflector and the Hilger 3-prism spectroscope. Of the four most conspicuous lines in the green, F is the most refrangible, and comparisons with burning magnesium showed one of them to be sensibly coincident with the edge of the magnesium fluting at 5006. The least refrangible of the four bright green lines was found to be slightly less refrangible than the carbon fluting near λ 517; it gave no indications of a fluted character, and further observations seemed to suggest that it was magnesium *b*, unless there were a very great change of position due to motion in the line of sight. The fourth line, which lies between F and 5006, is about one-third of the distance between them from F, and its wave-length, assuming the star to be at rest, was estimated to be about 490.

In addition to these, the G line of hydrogen was distinctly visible, and also a group of lines between G and F. The latter were not measured, as they appear on the photographs.

Amongst the fainter lines, one was estimated to be near λ 527, and was probably the iron line at E. By comparison with the spectrum of manganese chloride burning in a spirit-lamp flame, another line was found to be sensibly coincident with the edge of the brightest fluting of manganese, λ 557·6.

There was a bright line a little more refrangible than C, and a line at or near D was faintly visible.

Later on in the month eye observations were made on every possible occasion. The chief variations from those previously noted were the general fading of the continuous spectrum and the consequent unmasking of the lines between *b* and D. Micrometric measures of four new lines in this region were made by Mr. Fowler on February 23 and 24. These, with the other lines observed at Kensington in the region F to C, will be tabulated further on.

A light curve of the spectrum from F to C was drawn by Mr. Fowler and Dr. W. J. Lockyer on February 22, and confirmed by Mr. Fowler on February 23. The 3-foot reflector and McClean spectroscope were employed in each case.

The changes which took place in the Nova were exactly what

would be expected according to the hypothesis that new stars are produced by the collision of meteor-swarms. The rapid fading of the star demonstrated that small masses and not large ones were engaged, and this was further confirmed by the observed diminution in the brightness of the continuous spectrum relatively to the bright lines. If two condensed bodies were in collision, it is evident that the bright lines would fade first.

The Bright and Dark Lines.

A somewhat similar phenomenon to the bright and dark lines observed side by side on the photographs of the Nova had already been recorded by Professor Pickering, in the case of β Lyræ, and this has been confirmed by a series of photographs taken at Kensington. In this case, the bright lines are alternately more and less refrangible than the dark ones, with a period corresponding to the known period of variation in the light of the star. The maximum relative velocity indicated is stated by Professor Pickering as approximately 300 English miles per second.

In the case of Nova Aurigæ, however, the dark lines in all four photographs taken at Kensington were always more refrangible than the bright ones.

There was no evidence of revolution during the twenty days of observation. The relative velocity deduced from those of February 3, 7, 13 and 22 appears to be about 600 miles per second. As this only represents the velocity in the line of sight, we are still ignorant of the real velocities of the two bodies. The constant relative velocity indicated by the displacement of the bright and dark lines may be regarded as confirming the supposition that two meteor-swarms had collided, the velocities being so great, and the masses so small, that neither was captured by the other. On this supposition, the spectrum of Nova Aurigæ would suggest that a dense swarm was moving towards the earth with a great velocity, and passing through a sparser swarm, which was receding. The great agitation set up in the dense swarm would produce the dark line spectrum, while the sparser swarm would give the bright lines.

The relative velocity of 600 miles per second seems at first sight to be abnormally great, but, if we regard each of the component swarms as moving at the rate of 300 miles per second, the velocities are quite comparable with those of other bodies in space. The star 1830, Groombridge, for example, moves at the rate of 200 miles per second across the line of sight, and its velocity may be greater.

NOVA NORMÆ.

Another new star appeared in the southern constellation Norma in 1893. This was discovered on October 26, on a photograph taken at Arequipa, Peru, on July 10, 1893. Fortunately the photograph was one showing the spectra of stars instead of the simple images of the stars themselves, and the spectrum was seen to be identical with that of Nova Aurigæ.[1] Even more important were the observations of Campbell in February and March, 1894, when the star was about 10th magnitude. As the result of his work, he stated that "there can be no doubt that the spectrum of Nova Normæ is nebular."[2]

There can be little question, now that by modern methods we have secured permanent records of two Novæ, and find identical phenomena, that we are in possession of the main features of the problem presented by their appearance. Naturally more details are wanted, but the main points in their history are now garnered.

---

[1] *Ast. and Ast. Phys.*, 1894, p. 40.
[2] *Ibid.*, p. 312.

CHAPTER XIV.—HOW THE HYPOTHESIS HAS FARED.

WE are now in a position to bring together the results of the new observations made with instruments of modern research; to compare them with the early ones, and to see to what conclusions they lead.

In this chapter I shall have, in the first instance, to deal chiefly with the work and opinions of others along the various lines of inquiry opened up by the observations of the phenomena we are now considering.

When this has been done, we shall see how the new views have fared.

THE SPECTRUM OF NOVÆ.

For the purpose of seeking for the relationships of Novæ to the other celestial bodies I compiled a table in which the spectrum of Nova Aurigæ is compared with those of the nebula of Orion and the bright-line stars, and for the purpose of a more complete comparison I have added the observations of Nova Cygni, for the reason that it is not probable that two Novæ will arrive at identical maxima of temperatures, though it is quite certain they must end alike at a low temperature, and therefore give the nebular spectrum. The table is given in Appendix I.

The table indicates clearly that both in Nova Aurigæ and Nova Cygni we had to deal with a mixed spectrum; in fact, the addition of the lines of the bright-line stars to those observed in the nebula of Orion, together with others in the position of Duner's bands, almost build up the spectrum of the Nova.

The table also shows how small a basis of fact is at the disposal of those who consider that condensed bodies like our own sun can have any part to play in the phenomena of new stars.

From the beginning to the end of the action the principal lines of the nebulæ were seen both in Nova Cygni and Nova Aurigæ; at the end, as I have already stated, the nebular line was seen alone. It is important here to indicate that the change of intensity observed in the lines, as the light of the star wanes, is relative only, that is, to take one instance, the nebular line does not *become* brighter, it only *appears* brighter in consequence of the dimming in brightness of all the others as the intensity in the action going on is reduced.

THE RELATION BETWEEN NOVÆ AND VARIABLES LIKE MIRA CETI.

If the two swarms which produce a new star by collision are such that the mean distance between the meteorites in the resulting "mixed swarm" is about equal to that between the meteorites in a body of Group II, say α Orionis; mixed radiation of carbon and metallic fluting absorption will preponderate in the spectrum observed. At the same time, the sparser portions of the swarms will give us the radiation of the permanent gases. This was the state of things in Nova Coronæ, a detailed discussion of which has already been given. The general spectrum observed was one similar to that of α Orionis, but in addition, the presence of bright hydrogen lines was noted. This is a condition which cannot occur at any stage in the condensation of a single swarm, because a swarm dense enough to give the α Orionis type of spectrum would be too dense to give hydrogen and cleveite gas radiation. In a dense swarm absorption preponderates, whilst in sparse swarms radiation preponderates, the interspaces being flooded usually with incandescent hydrogen and helium.

In variables of the Mira type, we have almost a reproduction of the conditions of Nova Coronæ. As I suggested in 1888,[1] the variability in this class of stars is in all probability due to a swarm of meteorites revolving in a more or less elliptical orbit around a central swarm, the maximum luminosity occurring at periastron passage. At maximum, therefore, in such a variable, the luminosity proceeds from a mixed swarm, exactly in the same way as in a new star. At the maxima of Mira and other long-period variables, bright hydrogen lines have been observed, although the spectra are of the Group II type.

The Origin of the Mixed Spectrum of Novæ.

The discussion of the observations made of the changes that take place in the spectra of new stars, has already shown that the sequence of phenomena is strikingly similar to that which occurs in cometary spectra after perihelion passage. In general, however, there will be a difference: namely, that in comets there is usually only one swarm to be considered, whereas in new stars, there are two, which may or may not be equally dense. In new stars, we have accordingly the integration of two spectra, and the spectrum we see will depend upon the densities and relative velocities of the two swarms. At one part of the mixed swarm the temperature must generally be considerably higher than at another, in consequence of the greater number of collisions occurring locally, and the temperature will be lowest where the outliers are engaged.

---

[1] *Meteoritic Hypothesis*, p. 475.

In new stars, therefore, it is possible for us to have the radiation spectra of gases and vapours corresponding to sparse swarms of meteorites (nebulæ, bright-line "stars," and comets away from perihelion), and the mixed radiation of carbon, together with the fluting absorption of metallic vapours, corresponding to bodies of Group II, and to comets nearer perihelion. That is to say, we may have the radiation lines of hydrogen and cleveite gases and the fluting of magnesium at 500, *plus* the radiation of carbon, and fluting or line absorption of manganese, lead, iron, etc., a condition which cannot occur when only a single swarm is in question.

The mixed spectroscopic phenomena, which should be seen on the collision of two swarms, were noted in my paper of November, 1887,[1] as follows:—"We shall, in fact, have in one part the conditions represented in Class III*a* (Vogel), and in the other, such a condition as we get in γ Cassiopeiæ."

The Mixed Colours of Novæ.

It must be remembered that the sudden increase of temperature which determines the appearance of a new star is of quite a different character to the increase of temperature due to the condensation of a single meteoritic swarm. The phenomena accompanying each will therefore be different.

In the case of new stars, we have to begin with two meteoritic swarms, possibly in different stages of condensation. If no star or nebula were visible before, the sudden increase of light would be due to the collision of two undisturbed swarms or streams. If one of the swarms engaged already existed as a nebula, the collision of another with it would cause an outburst similar to that of Nova Andromedæ and Nova Aurigæ. If one swarm existed rather more condensed, the collision of another swarm with it would produce a higher temperature; this was

[1] *Proc. Roy. Soc.*, vol. xliii, p. 147.

the case with Nova Coronæ. But after the disturbance due to the collision had subsided, the temperature must begin to fall, as the mixed swarm is not in a condition to keep it up. *Novæ, therefore, run back along the temperature curve,* and their colour changes will in general take place in the opposite order to that followed by a condensing nebulous swarm. Hence the colour of new stars will be generally of a compound nature, and made up of the luminosity of the pre-existing nebula or star, together with the added light brought about by collisions.

The changes of colour will depend upon the relative amounts of light received from the two sources of luminosity of the Nova. This again may depend either upon the relative volumes occupied by the two swarms or upon their temperatures. If, for instance, the Nova consists of a large sparse swarm combined with a smaller and more condensed one, the light at first would be mainly that of the condensed part, the feeble blue or green light of the sparser swarm being overpowered. With rapid cooling in such a case, the light from the condensation would diminish in greater proportion than that from the larger mass, and the blue colour would then begin to assert itself. If the light from the condensation were reddish, this addition of blue light would tend to produce a purple tint, as in Nova Cygni, or a leaden one as in the star of 1572.

In Nova Coronæ the compounding of colours was very manifest.

12th May. White, with a bluish look. (Birmingham, *Monthly Notices,* vol. xxvi, p. 310.)

15th May. White, with a bluish look. (Baxendell, *Monthly Notices,* vol. xxvii, p. 5.)

16th May. Cream-coloured, *yellow seen through blue film. Monthly Notices,* vol. xxvii, p. 5.)

19th May. Buff-coloured. (*Monthly Notices,* vol. xxvii, p. 5.)

21st May. Leaden, slight orange tinge. (*Monthly Notices,* vol. xxvii, p. 5.)

22nd May. No yellow or red. (*Monthly Notices,* vol. xxvii, p. 5.)

23rd May. Dull grey. (*Monthly Notices,* vol. xxvii, p. 5.)

24th May.	Dull white, tinge of orange. (*Monthly Notices*, vol. xxvii, p. 5.)
25th May.	Slightly orange-white. (*Monthly Notices*, vol. xxvii, p. 5.)
26th May.	Dull orange-white. (*Monthly Notices*, vol. xxvii, p. 5.)
29th May.	Dull orange-yellow. (*Monthly Notices*, vol. xxvii, p. 5.)
2nd June.	Orange, no longer striking. (*Monthly Notices*, vol. xxvi, p. 298.)
25th June.	Orange-yellow. (*Monthly Notices*, vol. xxvii, p. 5.)
26th June.	Orange. (*Monthly Notices*, vol. xxvii, p. 5.)
11th July.	Dull yellow. (*Monthly Notices*, vol. xxvii, p. 5.)

The star had a yellowish tint to November 6, 1866. It appears, therefore, that when first visible the colour of this star was compounded of the yellowish-white colour of a swarm in an advanced stage of condensation and the bluish colour of a very early star, the blue colour in this case being due to the carbon radiation in the blue. This condition probably existed from May 12 to May 21. Afterwards, as the sparser swarm became very faint, the blue colour gradually died out, leaving a yellow tint preponderating.

The compounding of colours is, perhaps, more obvious in this case than in any other, and the reason is not far to seek, since the two swarms were of such very different degrees of condensation, one being comparatively far advanced along the temperature curve, whilst the other was only a very sparse swarm, as indicated also by the compound spectrum.

Dealing with all the colour changes chronicled, which are brought together in my communication to the Royal Society,[1] it will be seen that the changes of colour observed during the cooling of Novæ are perfectly in accordance with the sequence to which reference has been made. The Nova observed by Tycho Brahe appears to have reached an exceptionally high temperature, as indicated by its colour and brightness, and the changes of colour observed are exactly what they would have been in a cooling swarm of meteorites. This also was the case

[1] *Phil. Trans.*, vol. clxxxii, A, p. 440.

in the other Novæ of which colour observations have been recorded. Nova Cygni passed from a golden yellow to red, and then to orange, which agrees with the portion of the general colour sequence—reddish-yellow, yellow, red, yellowish-red.

Nova Andromedæ was first reddish-yellow, then orange coloured, reddish, and yellowish-red. Many observations of variations in the colour of Nova Coronæ were made, and these show that from bluish-white it ran down to a dull yellow. From these observations of colour it is evident that this was the hottest of the new stars. The "white, with a bluish look," recorded by Baxendell, is at the top of our colour stages, and the spectroscopic examination of the star indicated a high temperature. Baxendell noted that, after the first observations, no blue tinge was seen.

The discussion of colour observations therefore strengthens the view that new stars are complex bodies, probably produced by the collision of two swarms of different densities.

Carbon Radiation in Novæ.

In the *Meteoritic Hypothesis* I showed that the record of the presence of bright carbon flutings was unbroken from a planetary nebula through stars with bright-line spectra to those resembling α Herculis; that is entirely through Groups I and II of my classification. In comets, also, carbon is one of the chief features of the spectrum, and here there can be little doubt that we are dealing with swarms of meteorites.

Carbon is thus one of the chief characteristics of the spectra of uncondensed meteor-swarms. We have already seen that there is evidence of its existence in Novæ, but to emphasise this point, it may be convenient to summarise the observations which demonstrate it.

In Nova Coronæ the evidence depends upon two lines in the blue at approximately λ 467 and 473. The more refrangible was stated by Dr. Huggins to be either double or ill defined, and this was no doubt the maximum of luminosity of the compound carbon fluting at 468. The line at 473 was probably the least refrangible maximum of the same group. This double maximum of the carbon fluting has frequently been recorded in cometary spectra. The green flutings of carbon in Nova Coronæ were masked by the continuous spectrum, which, however, did not extend far enough into the blue to mask the other.

This double maximum was also seen by Vogel in Nova Cygni soon after its appearance, but after a time the sharp termination at 474 ceased to be visible, and the more refrangible one remained alone. Both these conditions have frequently been recorded in comets, and there is no doubt that they were due to carbon. In Nova Cygni, however, there was other evidence of carbon in the appearance of the brightest fluting at 517. Both this and the one at 468 faded away as the star gradually assumed the spectrum of a planetary nebula.

For Nova Andromedæ we have Copeland's statement that the spectrum was "the same as that given by any ordinary hydrocarbon flame." Although the flutings seen exist also in the nebula, it is probable that they were slightly intensified in the Nova, because the sáme observer did not note them in the nebula itself. Some of the observers, however, remarked that the spectrum of the Nova was that of the nebula intensified, although they did not recognise the true character of the nebula spectrum.

Again, a characteristic feature of the spectra of Novæ at some period is the apparent breaking up of the blue end of the spectrum into two parts. Thus, speaking of Nova Cygni, Vogel[1] says:—

> "It must be also mentioned as characteristic of this spectrum, that the blue and violet were very distinct compared with what they are in other stars possessing a band spectrum;"

[1] *Berlin Akad. Monatsber.*, 1877, p. 241.

and, with respect to an observation made on January 1, he further remarks:—

"After F followed a broad dark band which divided the spectrum into two parts."

A little later, he states that the dark band had a wave-length of 474—486.

Cornu[1] also showed the spectrum of Nova Cygni divided into two parts beyond F, and Dr. Lohse noted what he considered to be an intensely dark absorption band beyond F.

Again, Konkoly noted regarding Nova Andromedæ,[2]

"Shortly behind the region of F the spectrum is just as if cut off."

In all these cases, the apparent breaking-up of the spectrum was doubtless due to the existence of the bright blue fluting of carbon standing out beyond the end of the continuous spectrum, exactly as is seen in the spectra of bright-line "stars" and some of the condensing swarms of Group II.

In the discussion of the spectra of condensing swarms[3] I remarked:—

"When in these stars the spectrum is seen far in the blue, the luminosity really proceeds first from the carbon fluting, and, in the hotter stars, from the hydrocarbon one, which is still more refrangible, in addition. In the stars which have been examined so far, the dark parts of the spectrum, which at first sight appear due to absorption, are shown to be most likely caused by the defect of radiation in that part of the spectrum between the blue end of the continuous spectrum from the meteorites and the bright band of carbon."

In such cases as these just described the carbon fluting 468—474 appears broken off from the remainder of the spectrum.

Speaking of Lalande 13412, I wrote[4]:—

"The bright part of the spectrum extending from 473 towards the blue

[1] *Comptes Rendus*, vol. lxxxiii., p. 172.
[2] *Ast. Nach.*, No. 2681.
[3] Bakerian Lecture, 1888, p. 31.
[4] *Ibid.*, p. 35.

with its maximum at 468 is, I would again suggest, the carbon band appearing beyond the continuous spectrum;"

and again, referring to 2nd Cygnus:

"The bright band in the blue at 473 is most probably the carbon band appearing bright upon a faint continuous spectrum, this producing the apparent absorption from 486 to 473."

In 3rd Cygnus the same thing occurs, a dark band appearing from about 488 to 473, which is doubtless nothing more than the dark space between the end of the continuous spectrum and the carbon fluting at 474. I also remarked, concerning the origin of the discontinuous spectrum:—[1]

"I have already shown that when the meteorites are wide apart, though not at their widest, and there is no marked condensation, the spectrum will extend farther into the blue, and therefore the flutings in the blue will be quite bright; in fact, under this condition the chief light in this part of the spectrum, almost indeed the only light, will come from the bright carbon. Under this same condition the temperature of the meteorites will not be very high; there will, therefore, be little continuous spectrum to be absorbed in the red and yellow."

There can be no doubt, therefore, that the spectra of Novæ are similar to the spectra of bodies of Group II, and the later species of Group I, so far as carbon is concerned. All these, again, are closely related to cometary spectra, and the acknowledged meteoritic nature of the latter strengthens the view that Novæ are produced by the collisions of meteor-swarms.

THE EVIDENCE AS TO THE EXISTENCE OF TWO BODIES.

It will have been gathered from Chapter XI that in most of the earlier attempts which were made to explain the origin of new stars, the leading idea was that of a single body being suddenly disturbed in some way, with the possible result that the heat of its interior became manifested at the surface. Thus Zöllner, in 1865, suggested that the phenomena might be produced by the bursting of the crust which had just formed on

[1] Bakerian Lecture, 1888, p. 54.

the surface of a star approaching extinction. In connection with the new star in Corona, I pointed out in 1866 that all that seemed necessary to get such an outburst in our own sun was to increase the power of his convection currents, which we know to be ever at work. But a special subsequent study of the phenomena showed that this view was untenable, and I abandoned it, and in 1877 insisted upon the necessity for two bodies, the two bodies in question being considered as normally meteoritic swarms.

With regard to that same Nova of 1866, Dr. Huggins believed that the appearances were due to gaseous eruptions in a single body, and that—

"Possibly chemical actions between the erupted gases and the outer atmosphere of the star may have contributed to its sudden and transient splendour,"

which was likened to a world on fire.

Though Zöllner's theory was further advocated by Vogel and Lohse in 1877, the idea that such outbursts can be produced in a single body without external influence is now, I think, almost universally abandoned, though I should add that Dr. Vogel, in a modification of Zöllner's hypothesis, ascribes much of the increased luminosity of new stars to an actual "combustion" in one body.[1]

In relation to Nova Aurigæ the view of two bodies has gained considerable ground.

Dr. Vogel, who made some admirable observations during the appearance of this new star, states most distinctly that—

"We can no longer regard the assumption of a single body as sufficient in any explanation of the occurrence." [2]

Belopolsky also accepts two bodies,[3] and the views of Seeliger and Klinkerfues advocating two bodies have been much discussed.

---

[1] *Scheiner's Spectroscopy*, Frost, viii.
[2] *Ast. and Ast. Phys.*, 1894, p. 52.
[3] *Ibid.*, p. 54.

This then, so far as it goes, may be claimed as a victory for one contention of the meteoritic hypothesis; but, notwithstanding the general agreement as to the presence of at least two bodies in the outburst of Nova Aurigæ, there remain considerable differences of opinion as to the nature of the separate bodies, and of the kind of interaction between them.

Monck's view, referred to on p. 184, is the one extended by Professor Seeliger, who points out that the photographic investigations of Dr. Max Wolf and others leave but little doubt that space is filled with more or less extensive aggregations of thinly scattered matter, which may be called cosmical clouds, thereby accepting my view of a "meteoritic plenum," which was announced before any of these photographs were taken. He holds that if a heavenly body in rapid motion becomes involved in one of these cosmical clouds its surface will become heated, and the vapourised products will be partly detached and assume the velocity of the cloud; the fluctuations of brilliancy of a new star on this hypothesis are produced by the varying density of the cosmic cloud through which the body is passing.

I give two extracts from his communications:—

Recent photographic results—

"have left no doubt that space is filled with more or less extensive aggregations of thinly scattered matter."[1]

"It is from the nature of the case very probable that the supposed nebulous clouds or aggregations of dust-like particles, should be more numerous in certain parts of space than others."[2]

In this way he explains the observed fluctuations in brilliancy.

It is clear that Professor Seeliger accepts my general hypothesis of a meteoritic plenum, but not that part of it which explains many variable and all new stars by the clash of *two*

---

[1] *Ast. and Ast. Phys.*, 1892, p. 907.
[2] *Ibid.*, p. 917.

meteoritic swarms. But even this modified hypothesis of Professor Seeliger's has been strongly combated by Dr. Vogel.

The idea that such phenomena might be produced by the close approach of two bodies, and the consequent disturbances due to tidal action, was first started by Klinkerfues in 1865.

The tidal theory differs from Zöllner's only in ascribing the eruptions to the disturbances produced by tidal action when two bodies approach each other. This explanation, however, has met with much opposition on physical grounds.

Professor Seeliger remarks, in opposition to this view—[1]

"The static theory of the tides, which is used throughout, is quite incapable of giving a correct representation of the deformations which are doubtless produced by the close passage of the two bodies; for with very eccentric orbits (which it is necessary to assume on other grounds), the continually varying action would last for so short a time that one could scarcely expect to derive a trustworthy conclusion in regard to the actual circumstances from a consideration based on the forms which the bodies could assume in equilibrium."

Again, Vogel objects that—

"Sensible tidal action cannot be assumed to last for any considerable time, as on account of the great relative velocity of the bodies, they would separate at the rate of forty-six millions of miles per day."[2]

If we are to have tidal action let us have it under conditions with which we are familiar or which we can test by the light of experience. We know that the similarity between the spectra of the chromosphere and the nebulæ and the Novæ depends almost entirely upon the presence of two permanent gases, and we know also that if tidal action were set up in the sun, the body with which we are most familiar to-morrow, we should see no trace whatever of the spectrum of either.

Again, Mr. Maunder and others have pointed out that if the phenomena be due to tidal action producing the formation of solar prominences, the bright lines should be displaced to the

[1] *Ast. and Ast. Phys.*, 1892, p. 905.
[2] *Ibid.*, 1894, p. 54.

more refrangible sides of their normal places, for the reason that only those prominences on the side of the star presented to us would be able to produce visible bright lines, and such prominences would necessarily have their chief movement in a direction towards the earth. We have seen, however, that in Nova Aurigæ, the actual displacement of the bright lines was just the reverse.

The fact, moreover, that Nova Aurigæ ended by becoming a nebula is difficult to reconcile with the idea that in its earliest stages its luminosity was produced by outbursts of the nature of solar prominences. To have so-called "solar prominences," there must be a sun to produce them, and that must remain when the outbursts of the prominences has ceased; in this case the last stage of the spectrum of the new star should have resembled that of the sun. The fact that it did not indicates how worthless is the prominence suggestion in the light of modern knowledge.

Another very important objection to the solar prominence theory is this: If new stars are real stars capable of exhibiting prominence phenomena, then we have real stars ending as nebulæ, and thus clashing with the idea that nebulæ are "early evolutionary forms" of heavenly bodies. Further, if new stars be real stars, we should have to believe that the last expiring atmospheres of stars consist of hydrogen and unknown gases; but if we take the evidence afforded by the stars themselves we find that, instead of their last luminous atmosphere consisting of bright hydrogen and helium, the spectrum indicates the presence of absorbing carbon or carbon compounds.

These, however, are not the only objections which may be raised to the idea that we have to do with phenomena of the nature of solar prominences, whether produced by tidal action in the case of two bodies, or by a bursting of the crust which is forming in the case of a star approaching the end of its career as a luminous body.

We know of the existence of many bodies in space where all the known conditions of tidal action exist, bodies with fluid atmospheres, elliptic orbits, and not large perihelion distances, and yet in none of these have phenomena been observed at all approaching those presented by new stars.

But we may go further than this.

In the first place, there is no reason to suppose that the prominences in our own sun are produced by tidal action, and yet I suppose it was the spectroscopic evidence of apparent similarity between the Nova and the chromosphere which gave rise to the idea of tidal action.

In the second place, even if the fact that many of the lines seen in the spectrum of Nova Aurigæ during its first appearance were coincident with lines seen in the solar chromosphere, appears, at first sight, to support the idea, it will not do to forget that since the spectra of nebulæ also show chromospheric lines, the same argument might also be applied to prove that nebulæ are manifestations of prominences and tidal action. I do not imagine that very many will be prepared to believe that nebulæ are prominences, for if they are, they must be prominences of an unseen sun ! !

It is, I think, sufficiently evident, that each special hypothesis which has been brought forward to replace the view of the nature of the two bodies involved suggested by the meteoritic hypothesis has got no further than a damaging criticism from the authors of the others.

Dr. Huggins' Views.

I have found it undesirable in what has preceded to refer to Dr. Huggins' opinions on the various points involved, for the reason that all suggested explanations of the phenomena of new stars which have been put forward, except my own, have appeared to find favour with him in turn. I have thought it proper, therefore, to refer to them separately.

As I pointed out in Chapter XI, the appearance of the new star in 1866 was explained by him as a conflagration on the surface of the body; we were in presence of a world on fire.

In regard, however, to the more definite results secured by the observation of Nova Aurigæ, Dr. Huggins in the first instance found it necessary to suppose the existence of two bodies, in order to explain the phenomena observed.

Writing of the one body idea advocated in 1866,[1] he refers to it as

> "A view which, though not impossible, I should not now, with our present knowledge of the light changes of stars, be disposed to suggest."

He is then led to favour Klinkerfues' view of two bodies, the light variation being brought about, as I have shown, by tidal action. He refers to

> "Enormous eruptions, of the hotter matter from within, immensely greater, but similar in kind to solar eruptions,"[2]

thus brought into play. But although Dr. Huggins thus appeared to favour Klinkerfues' hypothesis, which is rejected by Seeliger, Vogel, and myself, he suggested another, which, so far as I can see, is diametrically opposed to it. His new special view assumes

> "Two gaseous bodies, or bodies with gaseous atmospheres, moving away from each other, after a near approach."[3]

Further, the gaseous bodies are not allowed to collide.

> "The phenomena of the new star scarcely permit us to suppose even a partial collision; though if the bodies were very diffuse, or the approach close enough, there may have been possibly some mutual interpenetration and mingling of the rarer gases near their boundaries."

On this hypothesis it need only be remarked that there is no evidence whatever that the "boundaries" of any star contain "rarer" gases. We know that in our own sun, where we can

---

[1] *Proc. Roy. Soc.*, vol. li, p. 495.
[2] *Ibid.*, p. 494.
[3] *Ibid.*, p. 495.

study such phenomena best, the "boundary" gives no indication of any gas, and there are good reasons why it should not do so. The gases hydrogen and helium involved in the phenomena with which Dr. Huggins deals in two mutually destructive hypotheses, exist *uncombined*, and spectroscopically visible only at the base of the solar atmosphere.

Finally, however, this idea was given up, and Dr. Huggins returned to the 1866 explanation, which I *think* has been abandoned by everybody else. I quote his words, written the next year:—

"Influenced by the analogy between some of the changes in the spectrum of the Nova, and those which are associated in the spectrum of β Lyræ with the variation of its light, and also by other reasons which we pointed out in our former communication, we are still strongly inclined to take the same view which we then ventured to suggest, namely, that in the outburst of the Nova we have not to do mainly with cold matter raised suddenly to a high temperature by a collision of any form, but rather, for the most part, as was suggested by Dr. Miller and myself in 1866, in the case of the first temporary star examined with the spectroscope, to an outburst of existing hot matter from the interior of the star or stars; indeed, to phenomena similar to, but on an immensely grander scale than those with which we are familiar in the periodic greater and lesser disturbances of the sun's surface.

"Such grand eruptions may well be expected to take place as stars cool, and if in two dull and comparatively cool stars such a state of things were imminent, then the tidal action due to their near approach might be amply adequate to determine, as by a trigger action, such eruptions.

"Under such conditions, fluctuations of brightness and subsequent partial renewals of the eruptive disturbances might well take place."[1]

THE FINAL STAGE OF NOVÆ.

In new stars the succession of events on the meteoritic hypothesis cannot be the same as that in the case of orderly condensation; there must be a *running backwards* of the phenomena.

We must have from the first appearance of a Nova to the

---

[1] *Ast. and Ast. Phys.*, 1893, p. 614.

last a "backwardation" ending in an "early evolutionary form." Increase of temperature is accompanied by spectral changes in a certain order; if the temperature is reduced the changes occur in reverse order, until finally we reach the "early evolutionary form" which the nebulæ are now acknowledged to be, even by the opponents of my views. This early form cannot be a mass of gas merely, because its temperature is lower than that of a sun, which it is potentially, and it must contain all the substances eventually to appear in the atmosphere of a sun.

After February the dimming of Nova Aurigæ placed the star beyond the reach of the Kensington instruments, but, as a matter of fact, the Nova reappeared in August, 1892, and was observed to have increased in brightness from the 16th magnitude in April to about 9th magnitude.

What, then, was the spectrum? It had almost completely changed; and among the first to observe the new spectrum was Professor Campbell, of the Lick Observatory. This observer then stated that "the spectrum resembles that of the planetary nebulæ."[1] In the following month the spectrum was also observed by Drs. Copeland and Lohse, and their observations seemed to them to "prove beyond doubt that Nova Aurigæ is now mainly shining as a luminous gas nebula."[2] The most striking evidence on this point, however, is that afforded by the photographic investigations of Von Gothard. He not only shows us the photographic spectrum of the new star at this stage of its history, but gives us also the spectra of several nebulæ to compare with it; and it is evident that we were certainly dealing, in the case of the Nova, with the same spectrum as in the nebulæ. Dr. Gothard, at least, was satisfied on this point, and stated that "the physical and chemical state

---

[1] *Ast. and Ast. Phys.*, 1892, p. 717.
[2] *Nature*, vol. xlvi, p. 464.

of the new star resembles at present (September and October, 1892) that of the planetary nebulæ." [1]

There was finally also telescopic evidence of the nebulous character of the stars.

Max Wolf wrote: [2]

"A number of new diffuse nebulæ were discovered in the vicinity of the star, and there even appeared to be traces of nebulous appendages proceeding from the star itself."

On my side then I may say, at all events, that I have the

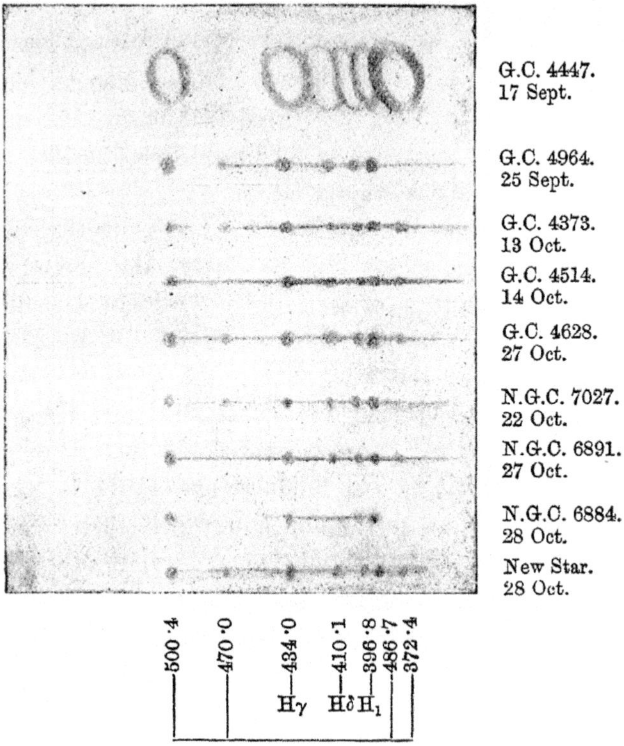

FIG. 69.—The spectrum of the new star in Aurigæ, as compared with the spectra of planetary nebulæ (Gothard).

[1] *Ast. and Ast. Phys.*, 1893, p. 55.
[2] *Ibid.*, p. 175.

great authority of the names of Campbell, Copeland, and Gothard, who state that they have certainly observed the spectrum to be that of the nebulæ, and of Max Wolf, who has photographed it.

This ending of the long series of spectral changes in a nebula has been the one which generally has caused the greatest surprise, but on the meteoritic hypothesis, this, and nothing else, must happen.

According to the general hypothesis, we have everywhere in space, as is now being abundantly revealed to us, especially by the photographs of Barnard, Max Wolf, and others, meteoritic aggregations, swarms, and streams, the constituents of which are, comparatively speaking, at rest, or are all moving one way, if they are moving at all, and undisturbed, because they are not being intersected by other streams or swarms at any one time.

But supposing any of these bodies cross each other, as unfortunately sometimes excursion trains cross each other, then there must be a change in the phenomena because there must be collisions; the collisions produce increased light, and we think that a new star is being born. Nothing of the kind. No new star is being born: there is simply a disturbance in a certain part of space.

Further, two sheets or streams of meteorites interpenetrating and thus causing collisions will produce luminosities which will indicate the condensation of each.

The spectra of the Novæ we are considering indicate that the colliding swarms were of different degrees of condensation, and the variations of light observed indicate several such encounters between less dense swarms after the most dense one had somewhat cooled down.

When the disturbance cools down we shall find that that part of space is still absolutely in the same order. In the case of Nova Aurigæ, and in the case of Nova Cygni after the war was over, nebulæ have been found to lie in the precise positions occupied by the new stars, and the only thing that one has to say about it is that the nebulæ were there before but that in consequence of our incomplete survey of the heavens they had not been observed.

After the new photographic chart of the heavens has been made, in future times, it will be found that all new stars are not really new, but the lighting up of something which existed there already. The argument for this view is simply this. If I light a fire, the smaller the fire the sooner will it go out, and the larger a fire the longer will it last. So if we are dealing in space with those illuminations which disappear in hours, days, or weeks, we cannot be dealing with any large mass; therefore the collisions in question cannot be between large masses of matter, but it must be a question of collisions amongst the smallest particles of matter.

It was abundantly clear then, that observations of a Nova with all the resources of modern science had established one of two main points of the meteoritic hypothesis, that we were dealing with sparse meteoritic swarms, which after the disturbance showed merely the spectrum of the nebula. There can be no doubt that a nebula really existed there before the disturbance.

It is interesting to consider one of the possibilities which may explain why small nebulæ may be overlooked in telescopic observations. In the so-called achromatic telescope, all the rays of light are not brought to quite the same focus, so that when ordinary stellar observations are being made, the focus is adjusted for yellow rays which are most luminous to the eye. Now the greater part of the visual light of a planetary nebula is confined to a single line of the spectrum in the green, so that the focus which is best adapted for observations of stars is not suitable for the observation of a small nebula, the nebula being out of focus, and its feeble light thus reduced by the diffusion of the image. This difference is much more marked in large than small telescopes, and Professor Campbell has pointed out that, on account of the steepness of colour curve in a large telescope,

> "A small nebula like Nova Aurigæ will, in general, appear relatively brighter in a small telescope than in a large one."[1]

On this hypothesis, then, we imagine a nebula in the position occupied by Nova Aurigæ not chronicled for the reason stated.

Whatever the density of the nebula may be, and it is not very sparse, it is undisturbed, it may even be a spiral nebula like the great one in Andromeda, and even in rapid rotation, so that when disturbed we get changes of wave-length according to the part that is pierced and changes of brilliancy if the various parts of it are not all of equal density.

This nebula is approaching us. It was disturbed by a much sparser stream rushing away from us, the relative velocity being over 500 miles a second. During the time of impact, the disturbances produced in the two swarms gave rise to a bright-line spectrum in the sparse swarm, and to a dark-line spectrum in the more condensed one. The spectrum of the sparse swarm disappears, the spectrum of the dense swarm changes gradually from dark to bright lines, and ultimately it puts on the original nebular spectrum. It is still there, and still approaching us.

General Conclusion.

Having thus endeavoured to bring together all the existing spectroscopic observations of Novæ, I now summarise the results. They indicate that some of the changes observed are closely related to those observed in cometary spectra, the difference in observing conditions, and the compound character of Novæ being duly allowed for. The temperature of a Nova depends upon the degree of condensation of the meteor-swarms

[1] *Ast. and Ast. Phys.*, 1893, p. 174.

which produce it. Its visibility depends also to a certain extent on its size. Hence it is that all Novæ do not attain the same maximum temperature or degree of visibility.

A more or less complete sequence of spectra has been found by considering the whole of the spectroscopic observations, and joining them together on a descending scale of temperature.

The general result of the comparison of new stars with variables, is that in passing from a variable to a new star, we pass from one swarm (or many) revolving in an elliptic orbit round a central swarm, to one probably revolving in a parabolic or hyperbolic orbit.

The changes in magnitude observed in Novæ are also in strict accordance with the meteoritic theory of their origin. The sudden increase of luminosity, which produces the appearance of a "new star," may well be due to the colliding of two swarms of meteorites, whilst the rapid fading away conclusively demonstrates, that small bodies and not large ones are engaged.

The complete discussion, therefore, tends to confirm the conclusion which I arrived at in November, 1887,[1] namely :—

"New stars, whether seen in connection with nebulæ or not, are produced by the clash of meteor-swarms, the bright lines seen being low temperature lines of elements, the spectra of which are most brilliant at a low stage of heat.

Replies to Objections.

We have next to consider the objections which have been urged against this hypothesis, and it will be convenient to deal at the same time with the various objections against the observations and conclusions of others which, as I have shown, have so strongly supported it. The objections are of a most trivial nature. One objection made by Vogel is that it is im-

---

[1] *Proc. Roy. Soc.*, vol. xliii, p. 154.

probable that the velocities could have been so great after collisions.

He wrote:—[1]

> "Nor is the question investigated, how the enormous relative velocity of over 460 miles per second can persist after the mutual penetration of two cosmical clouds or meteoric swarms, involving the close passage and inevitable collisions of particles whose masses are of the same order and the transformation of their energy of motion into heat."

In formulating this objection I think Dr. Vogel has overlooked the fact that high velocities after collision are more probable in the case of swarms of meteorites than in the case of formed stars.

On the meteoritic hypothesis we can escape from the difficulties produced by the old idea of collisions *en bloc*. Such objectors would urge that the velocity of a comet as a whole would be retarded by passing through the sun's corona, but we have instances to the contrary.

The motion of individuals only is arrested. The main body goes on.

Another objection has been raised by Dr. Vogel because, in relation to the Nova, I did not restate all I had previously written concerning the origin of bright- and dark-line spectra in stars.

> "Why all the particles of the denser swarm, or at least most of them, should give spectra with dark lines, and the particles of the sparse swarm for the most part, spectra with bright lines, is not further explained."[2]

I confess this objection amazes me, for it is one of the chief points of the meteoritic hypothesis that the main differences between bodies giving bright- and dark-line spectra is one of condensation only: a sparse swarm gives us bright lines because the number of meteorites in unit volume is small and the interspaces are great; a more condensed swarm gives us dark lines

---

[1] *Ast. and Ast. Phys.*, 1894, p. 53.

[2] *Ibid.*, p. 53.

because the number of meteorites in unit volume is greater, and the atmospheres of cooler vapour round each meteorite in collision begins to tell because the interspaces are reduced.

The following quotations will show how this matter stands:—

"If we assume a brightening of the meteor-swarm due to collisions as the cause of the so-called new stars, we have good grounds for supposing that in these bodies the phenomena should be mixed, for the reason that we should have in one part of the swarm a number of collisions probably of close meteorites, while among the outliers the collisions would be few. We shall, in fact, have in one part the conditions represented in Class III*a* (Vogel), and in the other such a condition as we get in γ Cassiopeiæ."[1]

"The discussion of the observations which have been made of the changes that take place in the spectra of new stars, has already shown that the sequence of phenomena is strikingly similar to that which occurs in cometary spectra after perihelion passage. In general, however, there will be a difference: namely, that in comets there is usually only one swarm to be considered, whereas in new stars, there are two, which may or may not be equally dense. In new stars, we have accordingly the integration of two spectra, and the spectrum we see will depend upon the densities and relative velocities of the two swarms."[2]

"The spectrum of Nova Aurigæ would suggest that a dense swarm is moving towards the earth with a great velocity, and passing through a sparser swarm, which is receding."[3]

I am the more justified in insisting upon the importance of this view that two bodies in different stages of condensation are involved, because, years after it was formulated by myself, Dr. Huggins apparently arrived at it independently—at all events he makes no reference to my prior announcements when he brings it forward as an explanation of the phenomena witnessed in Nova Aurigæ.

"The circumstance that the receding body emitted bright lines, while the one approaching us gave a continuous spectrum with broad absorption lines similar to a white star, may, perhaps, be accounted for by the two

---

[1] November, 1887. Lockyer, *Proc. Roy. Soc.*. vol. xliii, p. 147.

[2] November, 1890. Lockyer, *Phil. Trans.*, vol. clxxxii, A, p. 407.

[3] February 11, 1892. Lockyer, *Proc. Roy. Soc.*, vol. l, p. 435.

bodies being in different evolutionary stages, and consequently differing in diffuseness and temperature."[1]

Seeing that there are in the heavens thousands of permanent bodies with bright and thousands with dark lines in their spectra, I cannot imagine how it has been difficult for Dr. Vogel to understand how one (temporary) star should have bright lines in its spectrum, and another (temporary) star should have dark lines. All I can say is that upon such objectors lies the onus of producing a more simple (and yet sufficient) explanation than that I have suggested, and which it appears Dr. Huggins has also thought out independently.

I now approach an objection raised by Drs. Huggins and Scheiner relating to the spectrum in its latest stages. I have already shown that all the best observers, armed with the best instruments had recognised that the final spectrum of the Nova was a nebular spectrum with the lines at 500 and 495.

This is not the opinion of Dr. Huggins, who writes as follows :—[2]

"We wish to speak at present with great reserve, as our knowledge of the Nova is very incomplete ; but we do not regard the circumstance that the two groups of lines above described fall near the positions of the two principal nebular lines as sufficient to show any connection between the present physical state of the Nova and that of a nebula of the class which gives these lines."

To this Professor Campbell very naturally and promptly replied :—

"If the spectrum is *not conceded to be nebular*, I must ask what else we should expect to find in that spectrum if *it were nebular*."[3]

The answer to that is, that we could not expect to find anything else because it is all there. In fact, out of nineteen lines observed or photographed by Professor Campbell in the spec-

---

[1] May 16, 1892. Dr. Huggins, *Proc. Roy. Soc.*, vol. li, p. 494.
[2] *Ast. and Ast. Phys.*, 1893, p. 614.
[3] *Ibid.*, p. 727.

trum of the Nova, eighteen correspond perfectly with nebular lines. He adds:—[1]

> "Therefore the spectrum is nebular, and the fact that the lines have remained broad, or may have remained multiple, does not militate against the theory."

The telescopic and photographic evidence of the fact that Nova Aurigæ became a nebula, and Dr. Max Wolf's photographs of the Nova and its surroundings in 1893, resulting in the discovery of a number of new diffuse nebulæ in its vicinity, "and even traces of nebulous appendages proceeding from the star itself," are not replied to by Dr. Huggins.

Dr. Scheiner, although he accepts the theory of two bodies, explains the final nebular spectrum in a very original manner:—

> "Whatever may have been the circumstances of the blazing up of the Nova, it is at least certain that no less than two celestial bodies suffered an intense superficial heating, so that large masses of gas were thrown off from the bodies as if by an explosion; and these masses of gas, left behind the bodies, may be in such a condition as to give a spectrum like that of a nebula."[2]

It would be interesting to inquire what "gases" are in question, since two out of the three nebular lines are not acknowledged to be gaseous, and also what became of the known gases that must have been in the colliding bodies.

With reference to the drawing given by Dr. Huggins in support of his contention, two remarks may be made. The considerable broadening and perhaps even the multiple nature of the nebular lines observed by him are only the natural sequel to the phenomena presented by the star before the nebulous stage was reached. I have before pointed out many possible causes for changes of wave-length; and certainly the broadening of the many lines photographed at first, both bright and dark, were in all probability due to this cause.

---

[1] *Ast. and Ast. Phys.*, 1893, p. 729.

[2] *Scheiner's Spectroscopy*, Frost's Translation, p. 9.

CHAPTER XV.—THE NEW CLASSIFICATION OF STARS AND NEBULÆ.

THE EFFECT OF CONDENSATION.

THE next point in the meteoritic hypothesis—that some of the heavenly bodies are increasing, others diminishing their temperature—is one to which too great importance cannot be attached.

It has already been stated, with reference to the hypothesis of Kant and Laplace, and especially Laplace's view that in the nebulæ we have to deal, as also in the stars associated with them, with gases at a very high temperature, that on the meteoritic hypothesis I am compelled to differ in this particular both from Laplace and also from Vogel, who has most industriously attempted to establish a classification of the celestial bodies on the basis that they all are getting cooler.

I have already pointed out that in accordance with thermodynamical principles, the temperature must increase with condensation. A nebula condensing, then, must be a nebula getting hotter. We have already seen it demonstrated that the bright-line stars are bodies more condensed than nebulæ, consequently they will be hotter than nebulæ.

Professor Darwin has recently demonstrated that swarms of meteorites in space will behave exactly like a gas; therefore, what can be said of the thermodynamics of a gas may be said also of the thermodynamics of a meteoritic swarm, and if it be agreed that in accordance with dynamical theory, the temperature must increase with condensation so long as the conditions of a perfect gas hold good, and if we accept that a swarm of meteorites will behave like a perfect gas, then swarms of meteorites will also get hotter by condensation.

But in all such condensations as we are considering a time must arrive when the loss of heat by radiation will be greater than the gain due to condensation.

Then cooling begins.

As an example of a cooling body we have the sun. There are many quite independent lines of inquiry which show that that body was much hotter in past times than it is at present.

Now, it is pretty clear that the fundamental difference I have just pointed out must be largely taken into account in any valid system of classification. Of course we must classify our stars if we are to speak about them with intelligence, and understand the relations of one body or system of bodies to another.

The study of stellar spectra from the time of Rutherfurd to the present shows us that only a very small number of groups is in question. We seem to be in presence of an evolution in which only a very few variables are in operation, and in my opinion the phenomena suggest that the only variable of paramount importance is temperature.

In working out the classification of stellar spectra, which I communicated to the Royal Society in 1888, the course pursued was to study the flutings and lines of the various elements given in the existing lists and to fill up gaps in them by fresh experimental work with the view of finding the necessary criteria. The question, however, was complicated by the discovery in stellar spectra of many lines the origin of which could not be stated.

Some time has now elapsed since the classification was published. In the meanwhile the attempts to trace the origin of the unknown lines have been continued, and the discovery of terrestrial sources of helium and probably other gases, has thrown a flood of new light upon stellar chemistry.

I propose to trace the history of the criteria now at our disposal in the study of the phenomena of the stars, and to give the results of my latest researches.

EARLY CLASSIFICATIONS.

The new classification of stars which has been suggested by the totality of the facts and considerations which have so far occupied us is not the first classification of the stars by any means.

Although the first observations of stellar spectra were made by Fraunhofer, we owe to Rutherfurd the first attempt at classification. In December, 1862, he wrote as follows :—[1]

"The star spectra present such varieties that it is difficult to point out any mode of classification. For the present I divide them into three groups. First, those having many lines and bands and most nearly resembling the sun, viz., Capella, β Geminorum, α Orionis, Aldebaran, γ Leonis, Arcturus, and β Pegasi. These are all reddish or golden stars. The second group, of which Sirius is the type, presents spectra wholly unlike that of the sun, and are white stars. The third group, comprising α Virginis, Rigel, etc., are also white stars, but show no lines ; perhaps they contain no mineral substance, or are incandescent without flame.

"It is not my intention to hazard any conjecture based upon the foregoing observations ; this is more properly the province of the chemist, and a great accumulation of accurate data should be obtained before making the daring attempt to proclaim any of the constituent elements of the stars."

This classification was followed up by Secchi, who practically adopted Rutherfurd's three groups, changing, however, the word group to type, and adding a fourth. On this point Dr. Gould, in his memoir[2] of Rutherfurd, writes :—

"I cannot forbear calling attention to the classification, essentially the same, subsequently published by Secchi without reference to this or to

---

[1] *American Journal of Science*, vol. xxxv, p. 71.

[2] Read before the National Academy, April, 1895.

any of the other labours of Rutherfurd, and which is generally cited under Secchi's name." (See *Scheiner*, p. 258, and *Translation*, pp. 235–236.)

In these and other subsequent classifications it has been taken for granted that nebulæ have nothing whatever to do with stars, and that all the stars lie along one line of temperature, the highest temperature being at one end, and the lowest at the other; such, at all events, is Vogel's view. Now we have to consider that nebulæ are stars to be, and that some apparent stars are really nebulæ; and further, that the undisturbed nebulæ are of relatively low temperature; hence we have bodies getting hotter as well as bodies getting cooler, and both must be provided for.

In 1873 Dr. Vogel brought out a new and much more detailed classification considerably extending the number of groupings employed by Rutherfurd and Secchi. This classification is based on the assumption that all stars began by being very hot, and that the various changes observed in the spectra are due to cooling.[1] It was taken for granted, for some reason or other—possibly in view of the idea of Laplace—that all the stars in the heavens began in the condition of highest temperature, and that all that the stars did after that was to spend their millions and billions of years of life in getting colder; so that, if we could at the present moment find out which is the very hottest star in the heavens, we might be perfectly certain that every star in its beginning resembled it exactly in spectrum, and therefore in physical constitution: the presence of bright lines is considered as a matter of secondary importance only, and gives rise to sub-groupings only.

Dr. Scheiner has quite recently reiterated this statement. He appeals to his "new" observations of the spectrum of magnesium as a "direct proof of the correctness of the physical interpre-

[1] "Selon la théorie il faudra que tôt ou tard toutes les étoiles de la première classe deviennent de la seconde, et celles-ci de la troisième." Dunér.

tation of Vogel's spectral classes, according to which Class II is developed by cooling from I, and III by a further process of cooling from II.[1]

Pechüle was the first to object to Vogel's classification, mainly on the ground that Secchi's types 3 and 4 had been improperly brought together.[2] The views brought forward in support of the meteoritic hypothesis cut at the root of such a classification as this.

It is perhaps worth while in passing to point out that in 1886 I stated, taking the then classification as a basis:—[3]

> "On the nebular hypothesis, supposing . . . . that we started with ordinary cometary materials, then, on the beginning of a central condensation which in time is to become a star, as Kant and Laplace suggested, such central condensation should then give us a star of the fourth class. As the energy of condensation increased, and the temperature got higher, the spectra would change through the third and second classes, till ultimately, *when the temperature was highest*, the first class spectrum would be reached. *On the slackening down of the temperature* of the now formed star, the spectra of the second, third, and fourth classes would then be reproduced, but, of course, now in the direct order."

We now know that this classification will not do, since all reference to bodies with bright lines in their spectra is omitted; every one now, however, as I have shown, agrees that they must take the first place as representing "early evolutionary forms," and this is one of the teachings of the views I have been bringing forward for the last ten years.

The idea which one arrives at by a discussion of all the spectroscopic facts is that we begin with a condition in which meteorites in swarms and streams are very far apart, and from the collisions of these a spectrum results which gives us bright flutings and lines, in other words the spectrum of the nebulæ; when they become a little more dense, we get the

[1] *Ast. and Ast. Phys.*, 1894, p. 571.

[2] The details of Vogel's classification and Pechüle's criticisms are given in my *Meteoritic Hypothesis*, pp. 345–6.

[3] *Nature*, vol. xxxiv, 1886, p. 228.

bright-line stars; and as they become denser still, we find the stars with a mixture of bright and dark flutings. Then still more condensation and dark lines, and at length the highest temperature of all; after which begins a descent on the other side, till at last we end in cool, dark bodies like the earth and moon.

This seems to be the classification which is necessitated by the consideration of all the facts; and it is, moreover, one which appears to give us possibilities of an explanation of the phenomena of new stars and variable stars, and many other things without going into the region of the unknown and impossible.

TEMPERATURE-CURVE.

It also lands us in the so-called temperature-curve along which I ventured to place the various classes of nebulæ and stars some time ago. I am glad to say that so far no valid objection has been made to it.

In the first instance, minute differences were not in question. We can understand from the photographs given in Fig. 71 how perfectly justified Rutherfurd and others have been in attempting to classify the stars by means of their spectra. In Sirius we get one group of stars, distinguished by the development of certain lines, which are due to the absorption of hydrogen. In α Cygni the hydrogen is represented quite distinctly, but the absorption there with regard to certain lines is much more developed than in such a star as Sirius. In Arcturus the absorption of the hydrogen is almost hidden in an enormous mass of lines. Here again we have another large group, and it is not too early to remark that Arcturus in its spectrum exactly resembles our own sun. Thus we can say that, spectroscopically speaking, the sun is like Arcturus, not like α Herculis, α Cygni, and so on.

It will be noticed that in the classification I have suggested

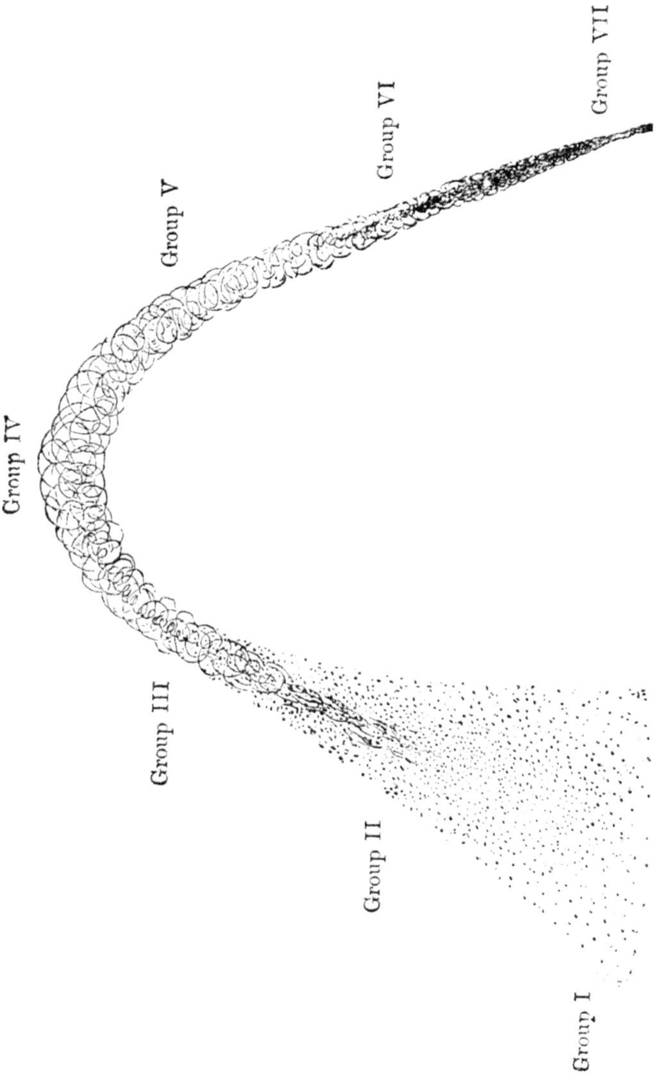

Fig. 70.—Temperature-curve.

I use the word "group," first employed by Rutherfurd; it is one which ought never to have been changed.

With regard to this subject, Professor Keeler agrees that a

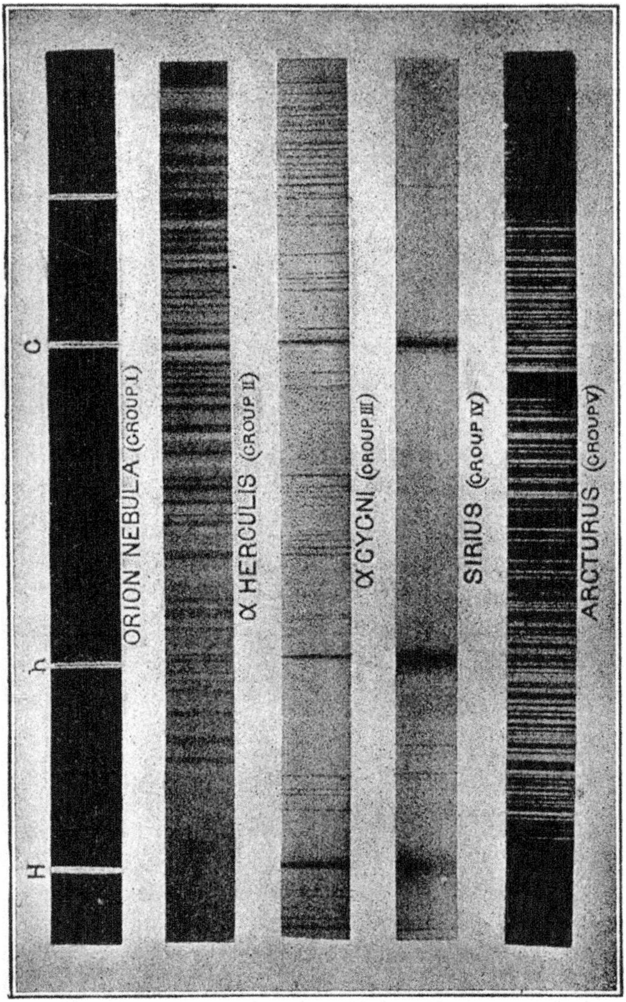

FIG. 71.—General diagram showing specimens of spectra in the first five groups.

classification which depends on this temperature curve has advantages over other systems. He writes :—[1]

"Professor Lockyer's system of stellar classification provides for both an ascending and a descending branch of the temperature curve, and in this

---

[1] *Ast. and Ast. Phys.*, 1894, p. 60.

respect it certainly has advantages over other systems which claim to have a rational basis."

Mr. Frost, the translator of Scheiner's book, writes as follows:—[1]

"Many spectroscopists are unwilling to admit that the only course of stellar development visible to us is along a line of descending temperature; in other words, they consider it quite as probable that the white stars represent the 'oldest' as the 'youngest' phase of stellar evolution. The further question may be raised—Is it not possible that a celestial body, after having reached the condition in which we define it as a star, may twice be red and exhibit a spectrum of the third type, once in its 'youth' and again in its 'old age'? In view of the fact that a gaseous body contracting under its own gravitation rises in temperature, until the laws of perfect gases cease to apply to it, we can hardly assert that a red star may not become hotter (under its own proportion of gaseous and liquid constituents) and finally pass into Class IIα or Iα."

On this paragraph I may remark that a reference to the names of the *many* spectroscopists who do not accept the idea of a line of descending temperature only, would have been very interesting.

Mr. Frost then goes to say:—

"Nevertheless, Vogel's view that we can observe only the descending branch of the temperature curve of stars appears to be confirmed by many of the more recent spectroscopic discoveries. We may cite the evidence afforded by the gaseous stars, by the intimate connection of nebulæ with stars of the first class, by the fact that the algol type variables—apparently young systems—belong to the first class, and by the relatively slight density of the binaries with spectra of Type Iα as compared with those of the solar class."

I have given the above extract in fairness to Mr. Frost, but I must confess I do not follow his line of reasoning. It seems, also, to contradict the preceding paragraph.

Professor Pickering's Classification.

In the classification of stars adopted by myself from a consideration of the visual observations in the first instance, only

[1] *Astronomical Spectroscopy*, p. 318.

the broader differences in the spectra were taken into account. Professor Pickering, however, has suggested a provisional classification in connection with the Henry Draper Memorial, for this purpose using the series of photographs of stellar spectra taken with small dispersion.

I am more glad than I can say that Professor Pickering, who has now given many years, with the aid of appliances beyond all precedent, to the study of these questions, has arrived at conclusions strikingly similar to my own.

In the first place he includes the nebulæ as well as the stars in his system; but it is right that I should add that he does not commit himself to any statements relating to the relative temperature of the different groups, although he distinctly accepts the idea of evolution, or what he terms "an order of growth."

He writes:—[1]

"In general, it may be stated that, with a few exceptions, all the stars may be arranged in a sequence, beginning with the planetary nebulæ, passing through the bright-line stars to the Orion stars, thence to the first type stars, and by insensible changes to the second and third type stars. The evidence that the same plan governs the construction of all parts of the visible universe is thus conclusive."

Professor Pickering's results may be shown in tabular form, but first it will be well to show the general differences between the more recent classifications:—

	Secchi.	Vogel.	Lockyer.
Nebulæ	Not classified	Not classified	Group I
Bright-line stars	"	Class Ic	
Mixed fluting stars	Type III	" IIIa	" II
Dark-line stars (ascending)	" II	" II	" III
Broad hydrogen stars	" I	" I	" IV
Solar stars (descending)	" II	" II	" V
Carbon absorption stars	" IV	" IIIb	" VI

[1] *Ast. and Ast. Phys.*, 1893, p. 722.

In his classification, Professor Pickering, unlike Vogel, includes what I consider as the earliest stages, taking the planetary nebulæ and such nebulæ as that of Orion into consideration; he then comes to the bright-line stars, that is, stars in which bright lines form the main feature, and then to such stars as those of Orion; and he ultimately places the Sun, as I also do, after such a star as Sirius.

There are two departures in his classification from that given by myself. One is that what I call the mixed-fluting group of stars, that is, stars with both bright and dark flutings, represented by several of the red, and brightest, stars in the heavens, he makes older than the Sun. And the class of stars which I group together and call Group VI. in which we get mainly the absorption of carbon in the atmosphere, he omits altogether, possibly for a very wise reason, as they are certainly the most difficult stars to tackle; but, after all, the divergencies in his classification from mine are small as compared with those between Dr. Vogel and myself, and Pickering, I repeat, like myself, attributes the variation to an "order of growth."

This premised, the differences of sequence between Professor Pickering and myself may be shown as follows:—

Lockyer.	Pickering.
I	I
II	
III	III
IV	IV
V	V
VI	II

Professor Pickering, in the Draper Catalogue, combines like stars under the different letters of the alphabet. The distribution of these letters in relation to my Groups is as follows:—

	Lockyer.	Pickering. (Draper catalogue.)
Nebulæ	I	P. (Planetary nebulæ)
Bright-line stars		O.
Mixed fluting stars	II	M.
Dark-line stars (ascending)	III	B. H. I. K (?).
Broad hydrogen stars	IV	A.
Solar stars	V	F. G. K. L.
Carbon absorption stars	VI	N.

It will be seen that certain groups are represented by more than one letter, but it is to be noted that here again Professor Pickering and myself have arrived at very nearly similar results, for generally a different letter with him represents a sub-group with me. This will be gathered from the subjoined table.

TABLE SHOWING THE SUBDIVISIONS OF GROUPS III AND V.

Group.	Pickering.
IIIα	H.
IIIβ	I. (some Q.)
IIIγ	B.
Vα	F.
Vβ	G.
Vγ	K. L.

With regard to Professor Pickering, then, I have chiefly to justify the place I have given to the stars of my Group II, which I place after the nebulæ and bright-line stars (as before defined), and he places after the Sun.

I fancy that one of the reasons which has led Professor Pickering to this conclusion is to be found in the assumption that strong indications of calcium and iron can only mark one stage of growth, while I think it is certain they must mark two.

We know they mark the present stage of the Sun's history, and taking meteorites as we find them, a relatively low temperature would provide us with more calcium and iron vapours to act as absorbers round each one than anything else.

Now we have strong indications of calcium and iron absorption in such stars as α Herculis as well as in the Sun, represented among the stars by Capella and Arcturus, but the general appearance of the spectra of these stars is so different that both Secchi and Vogel have classified them apart, and so indeed does Professor Pickering.

But the reason that I classified these stars also in different groups, and one on the rising and the other on the descending arm of the temperature curve, was that in those like α Herculis we have enormous variability as well as bright lines and flutings indicative of sparse swarms, while in those like the Sun the production of such phenomena is almost unthinkable. The special variability of stars of my Group II (Secchi's type III) and the production of bright lines at maximum is now freely acknowledged. On this point Professor Pickering remarks[1] :—

"Long period variables in general are of the third type, and have the hydrogen lines bright when near their maxima, as stated above. This property has led to the discovery of more than twenty objects of this class, and no exception has been found of a star having this spectrum whose light does not really vary. Of the variables of long period which have been discovered visually, the hydrogen lines have been photographed as bright in forty-one, the greater portion of the others being too faint or too red to be studied with our present means."

As said before, it seems impossible to imagine how our sun, as it proceeds along its "order of growth," should change into a body with such characteristics as these. But on this point we must wait for more large-scale photographic spectra; in other words, more facts.

Associated with this change in the order of evolution, Professor Pickering classes the chief stars in Orion, such as Bellatrix, characterised by spectra containing the dark lines of hydrogen and the cleveite gases, together with a few other dark lines of unknown origin, as early forms. On this point I may also quote the following from Professor Campbell (*Astronomy and Astro-Physics*, 1894, p. 475):—

---

[1] *Ast. and Ast. Phys.*, 1893, p. 721.

"In conclusion, I think we can say, from the foregoing observations, that the spectra of the Wolf-Rayet stars are not closely related to any other known type. They appear to have several points in common with the nebular and Orion type spectra; but the last two appear to be much more closely related to each other than to the Wolf-Rayet spectra. It is, therefore, difficult to place these stars between the nebulæ and Orion stars. They certainly do not come *after* the Orion stars, and one does not like to place them before the nebulæ. We can probably say that the bright lines are chromospheric, owing their origin to very extensive and highly heated atmospheres, but showing very little relation, in constitution and physical condition, to that of our own sun. For the present, at least, this type of spectrum must be considered as distinct from every other known type, just as the nebular spectrum is distinct, and like the nebular spectrum containing lines whose origin cannot now be assigned."

I shall show in the sequel the real cause of the great similarity of the spectra of the so-called Orion stars, and of the nebulæ as revealed by the new work. It is now possible to demonstrate that these bodies represent the extremes of celestial temperatures, and hence in any scheme of evolution they cannot be placed near each other.

DR. SCHEINER'S VIEWS.

Although Dr. Vogel and others apparently still hold in the main to the classification which assumes that all stars were created hot, and that nebulæ have nothing to do with them; that, in short, every star began in the highest stage of temperature, so that the whole history of every star in the heavens has been a process of cooling, there are signs of wavering here and there; some of the definitions are being changed to meet the facts which the photographic record is pouring in upon us. I may take, as an instance, the following statement made by Dr. Scheiner with reference to α Cygni, which is classified by Dr. Vogel as a solar star.[1]

---

[1] The "figures" referred to are micrometer measures of a photograph. My experience in these matters is that it is a pure waste of time to measure a

"These figures plainly show that the spectrum of α Cygni, in spite of the large number of its lines, has no resemblance with that of the sun. While it is possible to identify most of the lines with solar lines in respect to their position, yet the total lack of agreement as to intensity of the lines makes many of these identifications worthless." [1]

A comparison of enlarged photographs of the spectra of α Cygni and of the sun *which Dr. Vogel classes together*, shows at once the dissimilarity pointed out above without any measurement whatever. I am glad to find that Dr. Scheiner now regards the identification as "worthless," because it is such differences as these which have compelled me to reject Dr. Vogel's classification.

As I shall show, it is in relation to the stars last referred to, namely, those called the Orion stars, and others like α Cygni, that laboratory study of the spectrum effects produced at the highest temperature must be appealed to before any final conclusion may be drawn with regard to their ultimate classification; and, indeed, it is not too much to say that much of the work of the future, which eventually must smooth down all differences between stellar classifications, must consist of the study of single lines in the spectra of different stars.

---

photograph until it has been compared with others to which it is important to refer it, enlarged up to the same scale. In this I think I carry Professor Keeler with me (*Ast. and Ast. Phys.*, 1894, p. 485). "The coincidence of . . . lines is shown more beautifully by inspection of . . . . photographs than by any process of measurement."

[1] *Astronomical Spectroscopy*, Frost's translation, p. 247.

CHAPTER XVI.—THE NEW CLASSIFICATION TESTED BY PHOTOGRAPHS OF STELLAR SPECTRA.

THE OBJECT OF THE INQUIRY.

HAVING said so much on the different classifications of stars, and indicated, I trust judicially, that the one suggested by the meteoritic hypothesis has met with no serious objections, I now pass on to some recent work which was undertaken to test it by a limited photographic survey. In the first instance I had used the eye observations of others, for the reason that up to that time all my work had been chiefly directed to the sun.

So soon, however, as the research rendered it necessary to determine the sun's true place among the stars in regard to its temperature and physical conditions, arrangements were made to photograph the spectra of stars and nebulæ, in order to test the view, employing a quite new basis of facts. This new basis of inquiry consists of 443 photographs of 171 of the brighter stars.

My object was not so much to obtain photographs of the spectra of a large number of stars, as to study in detail the spectra of comparatively few; hence many of the stars have been photographed several times with special exposures and foci for different regions of the spectrum.

As in the case of stellar spectra observed by eye, the photographic spectra vary very considerably in passing from star to star.

Having this new and accurate basis of induction, the objects were to determine whether the hypothesis founded on eye observations is also demanded by the photographs, and in the affirmative case to discover and apply new tests of its validity or otherwise.

The Method of Grouping.

The basis upon which the first grouping of the photographs was founded was the extent of the continuous radiation at the blue end of the spectrum. Such a distinction was not possible in the case of the eye observations first discussed.

One of Kirchhoff's first conclusions in the infancy of spectrum analysis was that the hotter a light source the more its spectrum extends towards the violet and ultra-violet; and it is a fact of common knowledge that a white-hot poker is hotter than a red-hot one, the difference in colour being due to the addition of blue and violet light in the former case.

On this point I wrote as follows in 1892 :—

> " An erroneous idea with regard to the indications of the temperature of the stars has been held by those who have not considered the matter specially. It has been imagined that the presence of the series of hydrogen lines in the ultra-violet was of itself sufficient evidence of a very high temperature. The experiments of Cornu, however, have shown that the complete series of lines can be seen with an ordinary spark without jar. Hence the high temperature of such a star as Sirius is not indicated by the fact that its spectrum shows the whole series of hydrogen lines, *but by the fact that there is bright continuous radiation far in the ultra-violet.*"

We shall not go far wrong in supposing that the star with the most intense continuous radiation in the ultra-violet is the hottest, independently of absorbing conditions, which, in the absence of evidence to the contrary, we must assume to follow the same law in all.

A study of the stellar photographs taken at Kensington shows that there is a considerable variation in the distance to which the radiation extends in the ultra-violet. Some stellar spectra stretch far into the ultra-violet, while others leave off about the line K, and others again become dim between K and G.

Judged by this criterion alone, some of the hottest stars so far observed are γ Orionis, ζ Orionis, a Virginis, γ Pegasi, η Ursæ Majoris, and λ Tauri. Of stars of lower, but not much lower, temperature than the above, may be named Rigel, ζ Tauri, a Andromedæ, β Persei, a Pegasi, and β Tauri.

The first thing, then, to be done was to adopt this sorting process to all the photographs of stellar spectra I had available.

As a result of this preliminary sorting I obtained four marked groups; and when it was completed, I was in a position to consider the various divisions of the photographic spectra thus arrived at in relation to the groups which were previously suggested from a discussion of eye observations. It is clear that if I got the same results the first conclusions would be strengthened. Since I had succeeded in obtaining large dispersion photographs of the spectra, much detail was revealed, and hence I determined to deal with the presence or absence, or changes of intensity, of individual lines to a greater extent than Professor Pickering has done in his observations so far published.

Each of the four main groups determined by the length of spectrum in the violet was therefore next sub-divided into sub-groups by the most marked differences in the spectral lines. I do not propose to give the detailed inquiry in this place.

The First Results.

The important fact which stood out when the photographic attack had got so far was that, whether we took the varying thicknesses of the hydrogen lines or of the lines of other substances as the basis for the sub-arrangement of the spectra, it was not possible to place all the stars in one line of temperature, but it was necessary to arrange the stars in two series.

It must specially be borne in mind that the fundamental difference between other classifications and my own is that it demands the existence of bodies of increasing as well as bodies of decreasing temperatures. We have, therefore, to inquire how far this condition is satisfied by the mass of new facts at our disposal. This involves the consideration of some points in connection with the meteoritic hypothesis and a preliminary consideration of the probable effects of both increasing and decreasing temperatures.

Early in my researches I pointed out[1] that the absorption phenomena in stellar spectra need not be identical at the same mean temperature on the ascending and descending sides of the curve, since, on the meteoritic hypothesis, there must be a considerable difference in the physical conditions.

Still it was not to be expected that the differences in the line absorption would be very marked, and when I first suggested the new classification I fully recognised the difficulty of separating Groups III and V. Thus I wrote in 1888 :—

"With our present knowledge, it is very difficult to separate those stars the grouping of which is determined by line absorption into Groups III and V, for the reason that so far, seeing that only one line of temperature, and that a descending one, has been considered, no efforts have been made to establish the necessary criteria."

In the following year I gave the results of some visual observations of stellar spectra which seemed to justify the separation of the stars with line spectra into two groups, and to suggest the necessary criteria for distinguishing them.

In a condensing swarm the centre of which is undergoing meteoritic bombardment from all sides, there cannot be the equivalent of the solar chromosphere; the whole mass is made

[1] *Proc. Roy. Soc.*, vol. xliv, p. 26.

up of heterogeneous vapour at different temperatures, and moving with different velocities in different regions.

Some of the collisions may be end-on while others may be mere grazes, so that a mixed spectrum of high and low temperature lines might be expected.

In a condensed swarm, of which we can take the sun as a type, all action produced from without has practically ceased; we get relatively a quiet atmosphere and an orderly assortment of the vapours from top to bottom. But still, on the view that the differences in the spectra of the heavenly bodies chiefly represent differences in degree of condensation and temperature, there can be, *au fond*, no great chemical difference between bodies of increasing and bodies of decreasing temperature. Hence it is exceedingly probable that at equal mean temperatures on opposite sides of the temperature curve this chemical similarity of the absorbing vapours will result in many points of resemblance in the spectra.

This being premised, it is desirable to obtain a first approximation to the various phenomena *which may be expected to occur* under the different conditions of ascending and descending temperatures; considerations relating to the special chemical indications involved must be considered later.

The material involved in the discussion, according to the hypothesis we are dealing with, is meteoritic in its nature, and its temperature is low. This is the antithesis of Laplace's view, who held that it was gaseous and hot, and the first bodies considered, in which, owing to collisions, a condition of greater heat and visibility has been brought about, are the nebulæ and bright line stars. These constitute Group I in my classification. From these collisions, under the action of gravity, condensations will be produced.

Initially each pair of meteorites in collision may be regarded as a condensation.

Ultimately, when all the meteorites in the swarm which first appeared as a nebula are volatilised, there will only be one condensation, in the shape of a spherical mass of vapour. Between these points there must be other conditions.

I will take each of the groups *seriatim*.

Group I.—Nebulæ and Bright Line Stars.

Here we shall be dealing chiefly with interspaces, and, therefore, with gases, as opposed to metallic vapours.

The bright lines seen in the nebulæ should have three origins:—

(1) The lines of those substances which occupy the greatest volume (or largest area in a section); in other words, the lines of those substances which are driven furthest out from the meteorites and occupy the interspaces, when possibly they may be rendered luminous by electricity. These, of course, will be gases. With regard to the metallic lines the following assumptions may be made.

(2) The most numerous collisions between the meteorites in the swarm will be partial ones—grazes—sufficient only to produce slight rises in temperature. The nebular spectrum, so far as it is produced by this cause, will therefore depend upon the phenomena produced in greatest number, and we may therefore expect to find the low temperature lines of various metallic substances.

(3) In addition to the large number of partial collisions there may be a relatively small number of end-on collisions, producing very high temperature, and so far as this cause is concerned, there may be some metallic lines produced which are associated with very high temperatures.

On the hypothesis, the lines seen in the spectra of normal bright-line stars should, in the main, resemble those which appear in nebulæ. They will differ, however, for two reasons:—

(1) Owing to partial condensation of the swarm the gaseous area will be restricted, and the bright lines of gases will lose their prominence.

(2) On account of the increased number of collisions, more meteorites will be rendered incandescent, and the continuous spectrum will be brighter than in nebulæ.

Group II.—Stars with Bright and Dark Flutings.

At the stage of condensation following those of the nebulæ and bright line stars, the bright lines from the interspaces will be masked by corresponding dark ones produced by the absorption of the same vapours surrounding the incandescent meteorites. One part of the swarm will give bright lines, another dark lines at the same wave-lengths, and these lines will lose their importance in the spectrum. The interspaces will be restricted so that absorption phenomena will be in excess, and the first most obvious absorption will be that due to low temperature vapours, that is, fluting absorptions of various metals. Under these conditions we know from laboratory experiments[1] that the amount of continuous absorption at the blue end will be at a maximum. The radiation spectrum of the interspace will now be chiefly that of the gases evolved from meteorites.

The final condition of this stage will be that line absorption will become more developed, and the fluting absorption will decrease, both conditions being produced in different regions of action in the agitated swarm.

---

[1] Lockyer and Roberts-Austen, *Proc. Roy. Soc.*, 1875, p. 344.

Group III.—Stars of Increasing Temperature with Line Spectra.

With further condensation the radiation spectrum of the interspaces will disappear, and the fluting absorptions will be completely replaced by dark lines, for the reason that the incandescent meteorites will be surrounded by vapours produced at a higher temperature, the number of violent collisions per unit time and volume being now greatly increased. The line absorption and the continuous absorption at the blue end of the spectrum will diminish. The number of violent collisions per unit time and volume being still further increased, we should expect finally to deal only with those lines which have been gradually getting stronger, from whatever cause, during the previous stages.

It may be stated generally that, while the intense and irregular action is going on in various parts of the condensing mass, we may have effects which are much less likely to be produced after the action has ceased. We may get indications of many different temperatures, and the lines may be broadened, as they are in Novæ, in consequence of different velocities.

Group IV.—Stars of the Highest Temperature.

The stars with the simplest line absorption are collected together in Group IV. These stars will present to us the final result brought about by the successive stages of condensation. The continuous absorption in the violet will be at a minimum.

Group V.—Stars of Decreasing Temperature.

When we consider the cooling condition, that is, what happens when the temperature of the mass of vapour is no longer increased by the fall towards the centre of meteorites composing the initial swarm, we should expect to find the phenomena met with in stars with ascending temperatures presented *in the main* in the inverse order. These stars form Groups V and VI.

The lines seen broadest at the higher temperatures will continue to thin out. At an early stage the spectra will show the re-introduction of some of the lines which disappeared in the later stages in the ascending series.

The new lines will not necessarily be the same as those observed in connection with the stars of increasing temperature.[1] In the latter there will be the perpetual explosions of the meteorites affecting the special atmosphere of each, whereas in a cooling mass of vapour we have to deal with general phenomena.

With this increasing line absorption there will be a recurrence of the continuous absorption in the ultra-violet.

With the further thinning of the lines broadest at the highest temperature and reduction of temperature of the atmosphere the absorption flutings of compounds should come in.

So much, then, for what we should expect, assuming the hypothesis to be true. The result of the photographic survey quite justified these conclusions, and the remarkable facts stood out quite clearly—

(1) That the stars may be truly classified into two series, one of ascending, another of descending, temperatures.

(2) That at the highest temperatures in both series we deal chiefly with hydrogen and the Cleveite gases, as already stated in Chap. V, while at the lower temperatures on both sides we deal chiefly with low temperature arc and flame spectra of the metallic elements.

---

[1] *Proc. Roy. Soc.*, vol. xlv, p. 382.

Of the intermediate stars the knowledge was not so absolute, and obviously a further study was required. Many lines were classed as unknown, and the departures from the intensity of the lines in the spectra of the metallic elements were very marked.

As had been anticipated, there were some marked variations in the stellar spectra, which on other grounds appeared to be of about the same mean temperature on both arms of the curve.

CHAPTER XVII.—SOME NEW LABORATORY WORK.

ITS OBJECT.

I SHOWED in the last chapter that, although the photographic survey of stellar spectra which I had undertaken appeared to amply justify the classification which was based on the meteoritic hypothesis *in the main*, there were difficulties and uncertainties which required to be studied in the light of further work.

These difficulties had to do with the stars of intermediate temperature in which the intensities of some of the metallic lines were widely changed.

That the intensities of such lines could change had been long known, but, unfortunately for the present inquiry, dealing with the *photographic* region, the old observations had been made in the visible part of the spectrum chiefly and in relation to the sun, in which body changes of line intensity of the most remarkable character had been recognised and commented upon. It became of importance, therefore, to extend the observations of terrestrial spectra into the photographic region for the purpose of making the comparisons which were necessary for continuing the inquiry into the stellar spectra. Accordingly this work, which consisted of photographing metallic spectra at the highest available temperature, was next undertaken. In order to show the precise reason for it a few words by way of introduction are necessary.

LONG AND SHORT LINES.

As early as 1872[1] I showed that when an image of a light

[1] *Phil. Trans.*, 1873, vol. clxiii, Part 1, pp. 253—275.

source is thrown on the slit of a spectroscope the lines are seen of different lengths. The long lines represent the vapours which extend furthest from the centre of the light source, the short lines those which exist only at the centre. The adoption of this method of work enabled me to establish that when a metallic vapour is subjected at any one temperature to admixture with another gas or vapour, or to reduced pressure, its spectrum becomes simplified by the abstraction of the *shortest* lines as well as by the thinning of many *long* ones.

In another communication,[1] in 1873, I added that—

> "The test formerly relied on to decide the presence or absence of a metal in the sun (namely, the presence or absence of the brightest and strongest lines of the metal in question in the average solar spectrum) was not a final one ; and that the true test was the presence or absence of the *longest* line."

I gave a photograph of the spectrum of iron produced when an image of a horizontal voltaic *arc* was passed between iron poles and projected on to the vertical slit of a Steinheil spectroscope.

On the strength of the criterion thus established, I was enabled to announce the presence of many metallic elements in the sun's atmosphere not hitherto detected.

Some Short Lines indicate the Effects of High Temperatures.

It was generally assumed in the first instance that the short lines were true products of the greater heat of the centre of the arc.

Subsequent work with the jar spark, in 1876[2] and 1878[3],

---

[1] *Phil. Trans.*, 1874, vol. clxiv, Part 2, p. 490.
[2] *Proc. Roy. Soc.*, vol. xxiv, p. 352.
[3] *Ibid.*, vol. xxviii, p. 157.

showed that at spark temperatures some of the shorter arc lines behaved differently from others. Among these lines may be mentioned the two lines of calcium producing the solar lines H and K and two lines of iron at 4924·1 and 5018·6 (on Rowland's scale of wave-length).

These lines were enhanced in intensity on passing from the temperature of the arc to that of the spark, while many similar lines disappeared; it was found that the lines thus enhanced in intensity were of considerable importance in the spectrum of the solar chromosphere.

These facts seemed to show that short lines might be produced by two causes: (1) the increased temperature of the centre of the arc; (2) the rapid breaking up of the solid metal used as poles into various complex molecular groupings as the vapours passed from the core to the outer edge of the arc.[1] This last action was apparently responsible for by far the greater portion of the short lines.

Flame Arc, and Spark Lines.

In 1879 I attempted to carry the matter further by volatilising those substances which give us spectra in a Bunsen flame and passing a strong spark through the flame, first during the process of volatilisation, the substance being put into the flame just below the platinums; and then after the temperature of the flame has produced all the simplification it is capable of producing, the substance in this case being introduced into the base of the flame. The passage from flame to spark represented a stronger case than the passage from arc to spark, and the view that the above-named two causes were at work was greatly strengthened by the observations of magnesium, lithium, and many other metals. In the first place, the differences ob-

[1] *Proc. Roy. Soc.*, 1879, vol. xxx, p. 22.

served in dealing with different quantities were attributed to the fact that "the more there is to dissociate, the more time is required to run through the series, and the better the first stages are seen."

Further, lines invisible in the flame spectrum when the spark was not passing were rendered visible by the passage of the spark. The blue line of lithium about λ 4602 and the line of magnesium about λ 4481 may be given as examples. Some of the flame lines were dimmed or became invisible at the time of the production of the new lines. The lines intensified by the spark in the flame were the same as those enhanced on the passage from the arc to the spark.

This strengthened the view that the result of a higher temperature was to produce an important change in the spectrum, *and it was conceivable that in a space entirely heated up to the highest temperature the spectrum would consist entirely of the enhanced lines.*

The employment of the flame in the experiment just referred to suggested a series of observations at flame temperatures with a view of noting the difference between the flame and arc spectra. The oxyhydrogen flame as well as the Bunsen was employed, and observations of the spectrum were made when various metals were volatilised in the flame.

The general result of this line of work was to show that there was a step similar in kind to that from arc to spark between the flame and arc. To take iron as a case in point, a few lines only constitute the spectrum of the flame. The number is enormously increased on passing to the arc, but none of the flame lines are dropped.

By these and other experiments it was firmly established that the intensities of the lines common to the flame, arc and spark differ greatly at the three temperatures; so that the three stages of flame, arc, and spark lines are now generally recognised.

I next refer to one or two instances of the bearing of these results upon solar phenomena and the truly extraordinary variations observed when minute inquiries are made and *continued,* for "snap-shots" are quite useless in such investigations as these.

The differences may be shown either by the variations between the various spectrum phenomena, such as spots and prominences, and the ordinary solar spectrum, which is constant in spite of all local changes, or between the spot and prominence themselves in different regions and at different times; they are all set out in detail in my *Chemistry of the Sun.*

Let me begin with comparisons between the ordinary solar spectrum and that of the chromosphere.

In the visible region of the spectrum, iron is represented by nearly a thousand Fraunhofer lines; in the chromosphere it has only two representatives.

I showed in 1879 that there was no connection whatever between the spectra of calcium, barium, iron, and manganese and the chromosphere spectrum beyond certain coincidences of wave-length. The long lines seen in laboratory experiments are suppressed, and the feeble lines exalted. In the Fraunhofer spectrum the relative intensities of the lines are quite different from those of coincident lines in the chromosphere.

In sun spots we deal with one set of iron lines, in the chromosphere with another.

At the maximum sun-spot period the lines widened in spot spectra are nearly all unknown; at the minimum they are chiefly due to iron and other familiar substances.

The up-rush or down-rush of the so-called iron vapour is not registered equally by all the iron lines, as it should be on the non-dissociation hypothesis. Thus, as I first observed in 1880, while motion is sometimes shown by the change of refrangibility of some lines attributed to iron, other adjacent iron lines indicate a state of absolute rest.

It became obvious then that the change of intensity of spectral lines was a question of prime importance.

In a communication to the Royal Society in 1879, I pointed out that certain metallic lines were visible only at high temperatures, and that such work would help us in the study of "the atmospheres of the hottest stars."[1] In the same connection, in the *Chemistry of the Sun*, published in 1887, I gave the diagrams, here reproduced, indicating the lines of

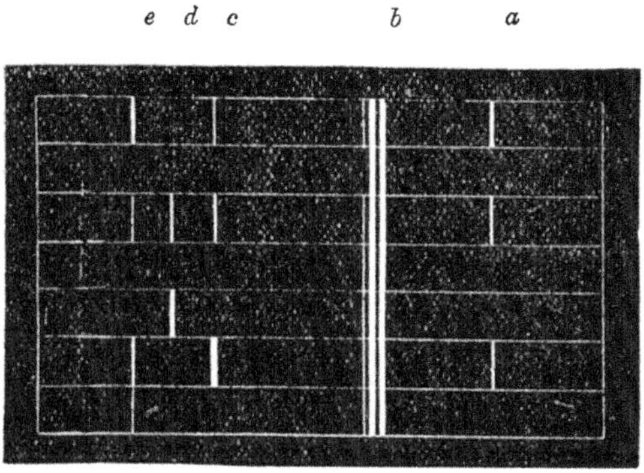

FIG. 72.—Showing the various intensities of the lines of magnesium as seen under different conditions.

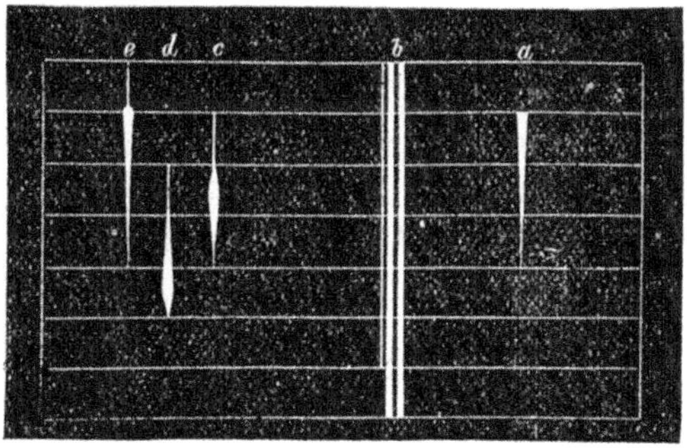

FIG. 73.—The various intensities of the lines of magnesium arranged in order of increasing temperatures. The lines marked *a*, *b*, *c*, *d*, *e*, in the diagrams have the following wave-lengths:—5209·8, 5167·5—5183·8 (*b* group), 4703·3, 4571·3, 4481·3.

[1] *Proc. Roy. Soc.*, vol. xxx, p. 22, 1879.

magnesium visible at various temperatures in the laboratory and in the sun and prominences.

I am rejoiced to find that the Potsdam observers are at length beginning to take this matter up. Dr. Scheiner has recently called attention to the behaviour of the line 4481·3 of magnesium, and agrees that the variations in the line observed are due to differences of temperature, and that therefore it may be used as a stellar thermometer.[1]

Had Dr. Scheiner been acquainted with my eighteen-years-old work, I am certain he would have done me the honour to quote, or at all events to refer to, it.

In the above diagram, Fig. 73, I indicate that the magnesium line at 4481·3 appears alone among the blue lines in the solar chromosphere.[2]

Details of the New Work.

The object of the new researches then was to see whether certain stellar anomalies in the photographic region followed suit with the solar anomalies which had been observed by eye.

For the new inquiries I have employed two storage cells giving a current of 7 amperes at 8 volts, with an Apps intensity coil giving a spark of 10 inches, and a jar capacity of about 0·03 microfarad.

The photographs were taken with an instrument having two prisms of 60°, a collimator of 3 inches aperture and 5 feet focus, and a photographic lens of 19 inches focal length. A Rowland

---

[1] *Astronomical Spectroscopy*, p. 8.

[2] Dr. Scheiner in referring to the same line in α Cygni writes as follows:—"The magnesium line at λ 4481 is the strongest in the entire spectrum. The other strong lines coincide for the most part with the fainter solar lines. The presence of numerous iron lines can be scarcely doubted, but here again we have the peculiar phenomenon that the fainter, instead of the stronger, lines occur. We may conclude from all these facts that very different conditions as to temperature must prevail in α Cygni from those in the stars of Class Ia."—Scheiner's *Astronomical Spectroscopy*, Frost's translation, p. 247.

grating of 21 feet 6 inches radius, with 20,000 lines to the inch has also been employed.

Iron.

I shall begin my account of the new work by dealing with iron.

Among the iron lines are two triplets, or sets of three lines, giving an example of repetitions of structure in different parts of the spectrum; one of them is less refrangible than G, and the other falls between *h* and H. In 1878 I referred to these as follows:—

"In many photographs in which iron has been compared with other bodies, and in others, again, in which iron has been photographed as existing in different degrees of impurity in other bodies, these triplets have been seen almost alone, and the relative intensity of them, as compared with the few remaining lines, is greatly changed. In this these photographs resemble one I took three years ago, in which a large coil and jar were employed instead of the arc. In this the triplet near G is very marked; the two adjacent lines more refrangible near it, which are seen nearly as strong as the triplet itself in some of the arc photographs I possess, are only very faintly visible, while dimmer still are seen the lines of the triplet between H and *h*."[1]

To clear the ground, it was important to determine whether the generally observed dropping out of lines in the spark depends upon the diminished quantity of incandescent vapour as compared with that in the arc.

With the brilliant spark obtained under the new conditions there is little difference between arc and spark with regard to the number of lines. Hence it may be concluded that the small number of lines previously recorded in the spark spectrum was not an effect of increased temperature, but one due to the small quantity of vapour produced by the use of a small coil.

The next point was to inquire if the photographic region of

[1] *Roy. Soc. Proc.*, vol. xxviii, p. 172.

the spectrum reveals lines which are the equivalents of those at 4924·1 and 5018·6 (Rowland's scale), which I had previously shown to be enhanced in passing from the arc to the spark spectrum,[1] and in relation to which several solar anomalies had been observed.

Seven additional lines were detected in the photographs at the following wave-lengths on Rowland's scale:—4233·3, 4508·5, 4515·4, 4520·4, 4522·7, 4549·6, 4584·0.

These have been confirmed by a reference to the map of the spark spectrum of iron published by Dr. McClean,[2] and attention was drawn to those at 4584·0 and 4233·3 in my paper on the arc spectrum of electrolytic iron.[3]

All these appear as short lines in the arc spectrum, so that the view that the short lines which appear in the arc spectrum can be divided into two categories, one including the lines which are brightened in the spark, and the other the lines which are not so enhanced is confirmed.

Having thus established that there are differences between the arc and spark spectrum, and that these differences are not due to the different quantities of vapour in the two cases, it must be concluded that a difference of temperature is the main cause of change.

Including the flame spectrum then, four distinct temperature stages are indicated by the varying spectrum of iron:—

(1) The flame spectrum, consisting of a few lines only, including the well-known triplets and many strong lines in the ultra-violet.

(2) The arc spectrum consisting, according to Rowland, of 2000 lines or more.

(3) The spark spectrum, differing from the arc spectrum in

---

[1] *Proc. Roy. Soc.*, vol. xxxii, p. 204.
[2] *Monthly Notices, R.A.S.*, vol. lii.
[3] *Phil. Trans.*, vol. clxxxv, A, pp. 995, 996.

the enhancement of some of the short lines and the reduced brightness of others.

(4) A spectrum consisting of the lines which are intensified in the spark. This we can conceive to be visible alone at the highest temperature in a space efficiently shielded from the action of all lower ones, since the enhanced lines behave like those of a metal when a compound of a metal is broken up by the action of heat.

A complete list of the iron lines seen at the different temperatures would be too long to reproduce here, so that the following statement of intensities is limited to the lines enhanced in the spark. The behaviour of these lines under the different conditions of experiment is as follows:—

LINES OF IRON WHICH ARE ENHANCED IN SPARK.

Wave-length. (Rowland).	Intensity in flame.	Intensity in arc (K & R) Max. = 10.	Length in arc (L) Max. = 10.	Intensity in spark (T). Max. = 10.	Intensity in hot spark (L). Max. = 10.
4233·3	—	1	—	—	4
4508·5	—	1	—	—	4
4515·4	—	1	—	—	4
4520·4	—	1	—	—	2
4522·7	—	1	3	—	4
4549·6	—	4	5	—	6
4584·0	—	2	4	—	7
4924·1	—	1	3	6	6
5018·6	—	4	—	—	6

K & R = Kayser and Runge, T = Thalén, L = Lockyer.

Calcium.

I next proceed to consider the results obtained in the case of calcium.

Among the chief observations of the spectrum of this metal are those made by Thalén, Kayser and Runge, and myself.

Thalén chiefly confined himself to a study of the spectrum at the temperature of the spark, the observations of Messrs. Kayser and Runge have been limited to the arc spectrum, while my own investigations have included all conditions of temperature available in laboratory experiments.

As I showed in 1876,[1] the most characteristic low temperature line is that at λ 4226·9, while the H and K lines are preeminent at high temperatures. The new work with the spark from the large intensity coil and large jars has shown that all the lines recorded by Kayser and Runge in the arc spectrum appear also in the spark spectrum, but with the exception of H and K, and two lines at wave-lengths 3706·18 and 3737·08, which do not appear to have been previously recorded in the spark, they appear with reduced relative intensities. The two ultra-violet lines are enormously enhanced in the spark.

As in the case of iron, four temperature steps can be recognised.

(1) The flame spectrum, in which the blue line 4226·9 is predominant, H and K and a few other lines being very feeble.

(2) The arc spectrum, in which the H and K lines are of about the same brightness as the blue line, while other feebler lines also appear.

(3) The spark spectrum, in which nearly all the lines of the arc spectrum are seen, but with reduced intensities, except in the case of H and K, which remain very bright, and two lines at 3706·18 and 3737·08, which are also very bright.

(4) A spectrum consisting of the two lines at 3706·18 and 3737·08, and the H and K lines, corresponding to a temperature higher than the average temperature of the spark, as before explained.

The complete spectra actually recorded are shown in the following table:—

---

[1] *Proc. Roy. Soc.*, vol. xxiv, p. 352.

CALCIUM.

Wave-length. (K and R.)	Intensity in flame (L). Max. = 10.	Intensity in arc (K & R). Max. = 10.	Length in arc (L). Max. = 10.	Intensity in spark (T). Max. = 10.	Intensity in hot spark (L). Max. = 10.
3706·18	—	4	—	—	8
3737·08	—	4	—	—	10
K 3933·83	3	10	10	10	10
3949·09	—	4	—	—	1
3957·23	—	6	2	—	1
H 3968·63	3	10	10	10	10
3973·89	—	6	4	—	2
4092·83	—	2	2	2	1
4095·25	—	2	—	2	trace
4098·82	—	4	4	2	1
4226·91	10	10	10	10	5
4238·00	—	—	2	—	—
4240·58	—	4	—	2	—
4283·16	—	8	6	8	2
4289·51	—	8	6	6	2
4299·14	—	6	6	6	2
4302·68	1	10	6	10	3
4307·91	—	8	6	6	1
4318·80	1	8	6	8	2
4355·41	—	6	4	—	1
4425·61	—	10	6	10	2
4435·13 }	1	10	} 8	10	} 3
4435·86 }	—	8		2	
4454·97 }	2	10	8	10	} 4
4456·08 }	—	8	} Out of range.	2	
4456·81 }	—	4		—	
4508·04	—	1		—	} 1
4509·89	—	1		—	
4512·73	—	1		—	1
4527·17	—	6		—	1
4578·82	—	8		4	1
4581·66	—	8		4	1
4586·12	—	10		4	2
4624·7	—	1		—	—
4685·40	—	4		—	—

K & R = Kayser and Runge, T = Thalén, L = Lockyer.

Magnesium.

Among other substances investigated in my earlier work was magnesium.

I showed in 1879[1] that in the flame spectrum the two less

[1] *Proc. Roy. Soc.*, vol. XXX, p. 22.

refrangible members of the *b* group were seen associated with a less refrangible line at 5210, making a triplet with them, while a line in the blue at wave-length 4571·3, and a series of flutings were also seen; on passing the spark, the blue line of the flame disappears, as well as the flame companion to *b*, while two new blue lines make their appearance at wave-lengths 4481·3 and 4703·3.

Among the spark lines is one at λ 4481, to which a brief reference was made in the preceding chapter. I first observed and thus described it in 1872:[1]

"This is a very brilliant winged line, but it appears short. Thalén makes it of the same intensity as the two at 4703·5 and 4586·5; but while this is excessively bright to me, 4703·5 is faint and 4586·5 invisible."

Taking *b* as having a length denoted by 4, I gave the length of this line at 4481 as 1: I also stated that it was not seen in the spectrum of the chloride, although the *b* group was distinctly seen. The line at 4481 has not to my knowledge been recorded by any observer as present in the arc spectrum, but a recent photograph shows it as a rather feeble line in the arc between poles consisting of magnesium. Using the large coil and jars, the line has also been photographed in the spectrum of the chloride; indeed all the lines recorded in the arc by Kayser and Runge have been photographed in the spark spectrum of the metal.

The work with the large jars has also resulted in the detection of another line of magnesium about wave-length 4395, which does not appear in the arc spectrum, and the line about 4587·4 observed by Thalén also shows itself feebly. The former of these is fairly bright and seems to be closely associated with 4481.

Again there are four distinct temperature steps, namely:

---

[1] *Phil. Trans.*, 1873, vol. clxiii, Part 1, p. 267.

(1) The flame spectrum, represented by lines at 4571·3 and *b*, a triplet in the ultra-violet, commencing with a line at 3734, and two flutings, one commencing at 5210 and the other at 5006·5.[1]

(2) The arc spectrum, comprising *b*, a line at 4352·18, and another triplet in the ultra-violet, commencing with 3838·4, 4481 being almost invisible, while 4395 and 4587 are quite invisible.

(3) The spark spectrum, including all the arc lines, but with 4481 intensely bright, 4395 fairly bright though short, and 4587·4 rather feeble.

(4) As 4481, 4395, and 4587·4 are much intensified in the spark spectrum, we can conceive a fourth stage at a still higher temperature, when magnesium would be represented by these lines alone. The complete spectra of magnesium under the three conditions at which observations can be made are indicated in the accompanying table.

[1] Messrs. Liveing and Dewar have ascribed these flutings to compounds of magnesium with hydrogen and oxygen respectively, but whether they are due to compounds or to the metal itself is immaterial for my present purpose.

MAGNESIUM.

Wave-length. (Rowland's scale.)	Intensity in flame (L). Max. = 10.	Intensity in arc (K & R). Max. = 10.	Length in arc (L). Max. = 10.	Intensity in spark (T). Max. = 10.	Intensity in hot spark (L). Max. = 10.
3720	8	—	—	—	—
3724	8	—	—	—	—
3730	6	—	—	—	—
3829·51	4	10	10	10 (C)	10
3832·46	6	10	10	10 (C)	10
3838·44	8	10	9	10 (C)	10
3850·2	—	4 (L)	—	4	2
3856·2	—	5 (L)	—	4 (H & A)	4
3892·7	—	3 (L)	—	4	4
3896·7	—	3 (L)	—	4 (H & A)	4
3987·08	—	2	—	—	2
4058·45	—	2	—	2 (L & D)	1
4167·81	—	1	—	—	4
4352·18	—	8	10	—	5
4395	—	—	—	—	3
4481·3	—	—	—	8	10
4571·33	7	4	1	2 (L & D)	2
4587·4	—	—	—	4	2
4703·33	—	8	6	8	4
4730·42	—	1	—	—	3
5006·5	10	—	—	—	—
b { 5167·55	—	8	3	8	7
b { 5172·87	10	10	5	9	9
b { 5183·84	10	10	5	10	10
5210	10	—	—	—	—

K & R = Kayser and Runge.
L = Lockyer.
T = Thalén.
C = Cornu.
H & A = Hartley and Adeney.
L & D = Liveing and Dewar.

CHAPTER XVIII.—APPLICATION OF THE NEW LABORATORY WORK TO STELLAR CLASSIFICATION.

METHOD EMPLOYED.

THE enhanced lines referred to in the last chapter have proved a real *open sesame* in the inquiry which now concerns us. It turns out that many of them have been formerly classed as unknown in stellar spectra, and that a complete study of all places in our hands a hitherto undreamt of instrument to be employed in minute stellar classification.

I am also glad to be able to say that its operation is most valid just in that region in which, as I pointed out in Chapter XV, further light was necessary to enable us to overcome difficulties and clear up uncertainties.

The photographic survey indicated that some of the principal stars might be arranged as follows, in order of ascending temperature, the hottest of them being placed at the top of the list,

Bellatrix,
ζ Tauri,
Rigel,
η Leonis,
α Cygni,
γ Cygni,
α Tauri,
α Orionis,

and that there were other cooling stars besides our sun, some of which could also be arranged in the following order of descending temperature:—[1]

Sirius,
β Arietis,
Procyon,
Capella,
Arcturus (Sun),

Arcturus is bracketed with the sun, because I have shown that its spectrum is like that of the sun, line for line.[2]

A still longer list of stars of descending temperature is as follows, Bellatrix, for the present, being taken as before as representative of the hottest stars:—

Bellatrix,
β Persei,
γ Lyræ,
Sirius,
Castor,
Procyon,
Arcturus.

Although the discussion specially deals with the hotter stars it is necessary to include references to the cooler ones in order to contrast the behaviour of the high temperature lines with those which are characteristic of low temperatures, and further, to compare the appearances of the lines at different stellar temperatures.

α Orionis is taken as a typical case of a relatively cool star which is bright enough to be studied with sufficient precision for our present purpose.

The way in which I have attempted to use the new information is the following:—A special examination of the photographs of the selected list of stars has been made in relation

---

[1] *Phil. Trans.*, vol. clxxxiv, A, p. 709.
[2] *Ibid.*, p. 699.

to the lines of the three metals first inquired into, namely, iron, calcium, and magnesium, and maps have been drawn to indicate the behaviour of the enhanced lines in each stellar spectrum. In the maps the lines photographed in the spectrum of each star are drawn with lengths proportional to their estimated intensities, the strongest line in each being represented by a line equal to the full width of the strip devoted to each spectrum.

ASCENDING TEMPERATURES.

Iron.

I began with iron, as it is the substance to which I have given the longest study. The accompanying diagram (Fig. 74) shows the results obtained:—

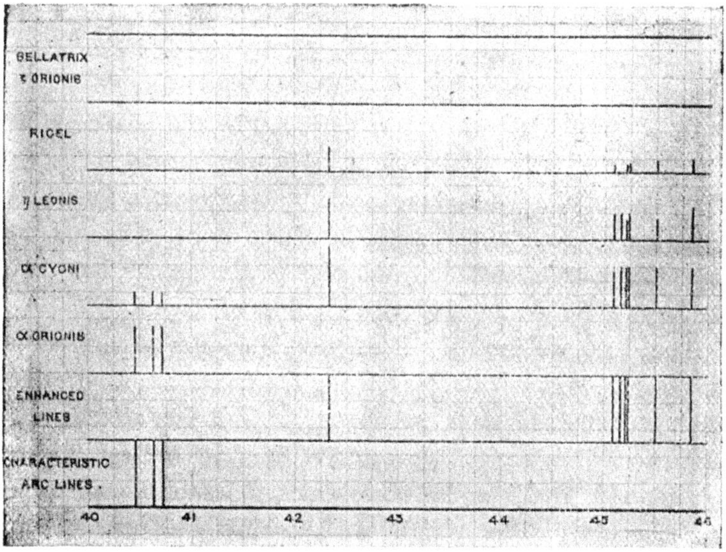

FIG. 74—Behaviour of iron lines in stars of increasing temperature.

The various results recorded in the map may be stated as follows:—

In α Orionis we have iron represented by low temperature lines corresponding to a temperature intermediate between the first and second stages.

Only two of its more prominent lines *may* coincide with enhanced lines in the region compared. As there are so many lines in the spectrum of this star, it is possible that these may be only chance coincidences, and this is the more probable since the lines which are most enhanced are certainly absent. Further, I have already shown[1] that

> "The temperature of the most important iron-absorbing region in α Orionis is nearer that of the oxy-coal-gas flame than that of the electric arc; it is probable that the average temperature is intermediate between that of the arc and that of the flame, but nearer to the latter."

Passing to γ Cygni, the iron lines confirm the idea that the temperature is higher than in α Orionis. Here the enhanced lines of iron are stronger than the arc lines representative of the second stage of temperature; that is, the temperature is approaching the fourth stage.

In α Cygni, in which star, to judge by the extension of the spectrum towards the violet, the temperature is higher, the enhanced lines of iron are among the strongest in the whole spectrum. At the same time, some of the stronger lines of the stage 2 spectrum, including the triplets, also appear, but they are very feeble as compared with the lines of stage 4.

Here, then, we probably have absorbing iron vapour at a temperature very nearly approaching the fourth stage. If this result be confirmed, we at once get an explanation of the great differences of intensity between the lines in α Cygni which coincide with iron lines, and those which appear in the spectrum of iron as observed terrestrially with an unshielded arc.

η Leonis represents a stage of temperature a little higher than that of α Cygni; here the second stage iron lines have disappeared altogether, and the enhanced lines appear alone.

At the higher temperature of stars like Rigel, the enhanced

[1] *Phil. Trans.*, 1893, vol. clxxxiv, A, p. 703.

iron lines also appear without any trace of the familiar iron spectrum typified by the triplets, and there seems to be little doubt that we are here in presence of iron vapour at a transcendental temperature corresponding to the fourth stage to which I have drawn attention.

In ζ Tauri, the iron lines are almost identical with those in Rigel.

At the still higher temperature of Bellatrix, all visible traces of the iron spectrum have vanished from the photographs.

GENERAL RESULTS WITH REGARD TO IRON.

The general result of the investigation of the enhanced iron lines in stellar spectra confirms the view that the absorbing regions of the hottest stars exist at a higher temperature than any attainable in laboratory experiments, the spectrum of iron consisting solely of those lines which are enhanced in passing from the arc to the spark. At the same time, some of the lines in the spectra of the hottest stars formerly classed as unknown are now shown to be due to iron.

The fact that even the enhanced lines themselves disappear from the spectra of stars approaching, or at, the acme of stellar temperature raises another question to which reference will be made later.

It is obvious that the enhanced lines may be absent from the spectrum of a star, either on account of too low or too high a temperature. In the case of a low temperature, however, iron is represented among the lines in the spectrum, but at the highest temperature all visible indications of its presence seem to have vanished. This result affords a valuable confirmation of my view that the arc spectrum of the metallic elements is produced by molecules of different complexities, and it also indicates that the temperature of the hottest stars is sufficient to produce simplifications beyond those which have been produced in our laboratories.

The facts which are graphically represented in Fig. 74 indicate that, so far as these stars are concerned, the results, with regard to stellar temperatures determined by a study of the iron lines, are identical with those to be gathered from the extension of the radiation spectrum into the violet. Independently of the extensions of spectrum into the violet for different stars, then, the relative temperatures may be determined by a study of the iron lines. Thus, a star, in the spectrum of which iron is represented by traces of the triplets characteristic of the arc spectrum as well as by the enhanced lines, must be cooler, so far as the absorbing iron vapour is concerned, than one in which iron is represented by the enhanced lines alone. Similarly, we must conclude that a star in which iron has no representative lines is hotter than one in which the enhanced lines appear without the arc lines. In practice the iron lines furnish a much more convenient indication of stellar temperature than the continuous spectrum, for the reason that in the case of iron no special photographs are necessary, while for an investigation of the continuous spectrum special photographs of stars with very carefully controlled exposures have to be taken.

CALCIUM.

The calcium lines in the spectrum of α Orionis indicate that we have in that case a temperature not greatly differing from that of the second stage; the blue line which is characteristic of low temperature, as well as H and K being very strongly developed.

The behaviour of the calcium lines is indicated in Fig. 75.

On passing to the stars of successively higher temperatures which have already been studied in the case of iron, it is seen that the blue line has disappeared at the temperature of α Cygni, while the H and K lines persist with gradually reduced intensities up to the hottest stars. The intensities of the H and K lines as compared with the blue line fully bear out the results as to the order of temperature of the stars which has been derived from a comparison of the continuous radiation spectra, and also from the appearances of the iron lines. Unfortunately, the Kensington photographs of stellar

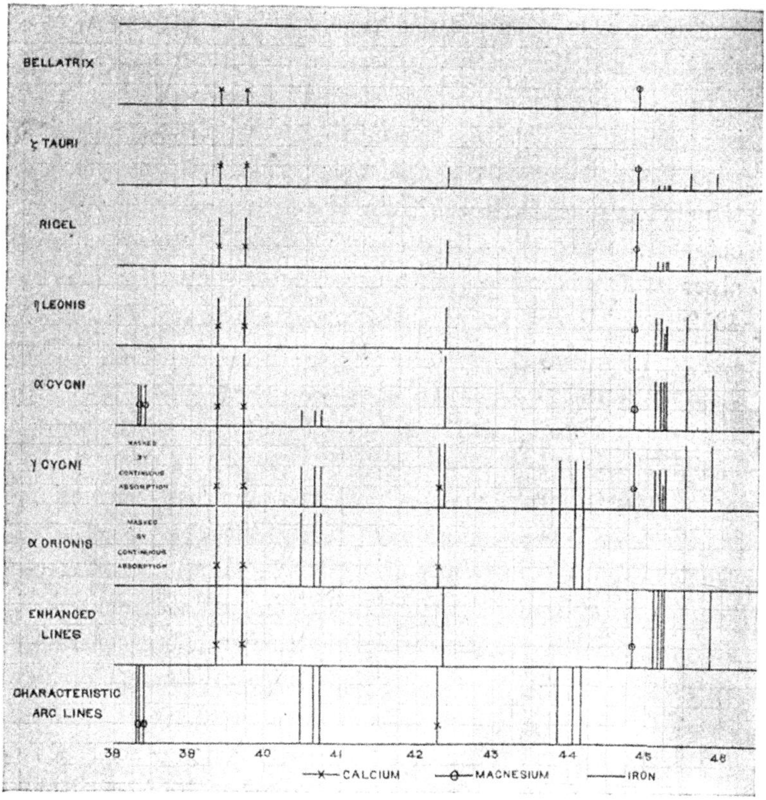

FIG. 75.—Behaviour of calcium and magnesium lines in stars of increasing temperature.

spectra which have so far been obtained do not extend far enough into the ultra-violet to permit a complete investigation of the varying appearances of the two ultra-violet enhanced lines referred to in the last chapter. Both of them appear in the spectrum of Sirius, however, which is certainly a hot star though not one of the hottest.

Magnesium.

Magnesium furnishes us with very definite indications of the four stages of temperature, and the discussion of its representative lines in stellar spectra is therefore very important.

In α Orionis, the *b* group is strongly developed, while 4481 is absent. In Fig. 75, the ultra-violet triplet commencing with 3838 is taken as typical of the second stage spectrum, but the photographs of the spectrum of α Orionis which have been obtained do not extend far enough to enable the presence or absence of this triplet to be ascertained. Another line, at 4352·2, which appears at the second stage of temperature is, however, probably present in α Orionis. The flame line 4571·3 is also present,[1] and I have also shown the probable presence of the two flutings of magnesium in the spectrum of α Orionis and similar stars.[2]

In α Orionis, then, the most effective absorbing magnesium vapour is at a temperature not greatly different from that of the flame, but the presence of 4352·2 indicates that the temperature must be slightly higher. It is important to remember, however, that on the meteoritic hypothesis, different parts of such a swarm as α Orionis may have widely different temperatures.

Taking γ Cygni, as before, to be considerably hotter than α Orionis, as indicated not only by the length of continuous spectrum but by the iron and calcium results, there is a considerable change in the lines representative of magnesium. The flutings have quite disappeared, and the spark line at 4481 appears as a well-marked line. Some of the arc lines, includ-

---

1 *Phil. Trans.*, 1893, vol. clxxxiv, A, Plate 28.

2 *Proc. Roy. Soc.*, vol. xliv, p. 54.

ing 4352·2 and 4167·8, remain, so that in the absence of a record of the ultra-violet triplet, it may be concluded that the absorbing magnesium in γ Cygni is probably at a temperature not differing greatly from that of the spark, that is, the third stage of temperature. The wave-length of the new spark line about 4395 is not yet known with sufficient accuracy, on account of its great breadth, to justify its use in this inquiry. The actual appearance of the magnesium lines in γ Cygni thus confirms the conclusion with regard to the temperature of this star which has already been derived from the discussion of the lines of calcium and iron.

In the case of α Cygni the most prominent magnesium line is the spark line at 4481. The *b* lines, the ultra-violet triplet commencing with 3838·4, and the line 4352·2 are also present, while 4571·3 and 4167·8 are absent, or very feeble. The great intensity of 4481, which is only a short line in the spark, indicates that the temperature is, in all probability, a little higher than that of the experimental spark, that is, intermediate between the third and fourth stages.

Passing to η Leonis, 4481 is a little less intense than in α Cygni, while the line at 4352 is considerably reduced in intensity as compared with α Cygni. A temperature a little higher than that of α Cygni is therefore indicated.

In Rigel, where the temperature is higher than in η Leonis, 4481 is one of the few strong lines recorded in the spectrum, and it appears without the other magnesium lines.

The same is true of ζ Tauri and Bellatrix, except that 4481 is now reduced in intensity. These varying appearances are indicated in Fig. 74.

The study of magnesium thus perfectly accords with what we have learned as to relative stellar temperatures from a discussion of the lines of iron and calcium.

General Results with regard to Calcium and Magnesium.

Fig. 75 indicates that in the case of the stars so far discussed, those namely with ascending temperatures, the same order of temperature is arrived at by a consideration of the lines of calcium and magnesium as that deduced in the first instance from the relative lengths of continuous spectrum, and afterwards by an inquiry into the presence of the enhanced iron lines. Four indications of stellar temperatures are therefore now available, namely, the extent of the continuous radiation, the lines of iron, the lines of calcium, and the lines of magnesium.

The enhanced lines of calcium and magnesium, unlike those of iron, do not disappear from the spectrum in the case of the hottest stars yet studied, but they become very feeble, so that an approach to disappearance is indicated.

Descending Temperatures.

I now pass to the discussion of those stars which the previous investigations indicated to be cooling bodies.

Fig. 76 indicates the variations of the metallic lines in this series of stars. At the highest temperature magnesium and calcium are represented by their enhanced lines alone; with a fall of temperature the enhanced lines of iron are next added, and, later, the arc lines of all three metals; at the lowest temperature, the enhanced line of magnesium disappears, while those of iron are either absent or very weak; the enhanced lines of calcium, on the other hand, which are relatively strong in the arc, remain.

The facts recorded in the photographs of each star are as follows:—

Starting with Bellatrix, magnesium is solely represented by the enhanced line at 4481, which is relatively feeble, calcium by the H and K lines, also feeble, while there are no indications of iron at all.

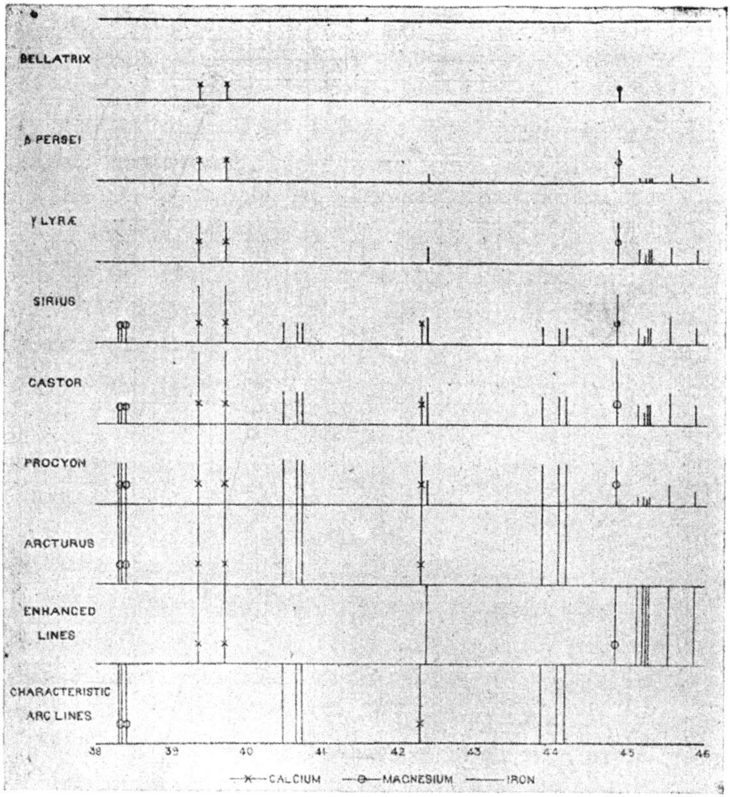

FIG. 76.—Behaviour of metallic lines in stars of decreasing temperature.

Passing next to *β* Persei, 4481 and the H and K lines are intensified, while the enhanced lines of iron are added.

In *γ* Lyræ, all the enhanced lines of magnesium, calcium, and iron are strengthened.

A further intensification of the enhanced lines occurs on passing to Sirius, but here the characteristic arc lines of all three substances are added, indicating a temperature intermediate between that of the experimental spark and that at which the enhanced lines appear alone.

With a further slight fall of temperature, represented by Castor, the enhanced lines become slightly stronger, as do also the characteristic arc lines.

At the still lower temperature of Procyon there is a marked increase in the intensity of the characteristic arc lines and a decrease in the intensity of the enhanced lines, except in the case of H and K, which, as already pointed out, are strong lines in the arc.

Finally, at the temperature of Arcturus, or the sun, the enhanced lines of magnesium and iron are either very faint or entirely absent, while H and K remain visible for the reason already stated, and the arc lines of all three metals become very strong.

The Lines of Gases in Stellar Spectra.

The maps, I think, will indicate the considerable advance which has been made with regard to the lines in stellar spectra, but the inquiry is not yet complete.

So far I have only considered the lines of certain metallic elements, and their comparison with stellar lines. It is now important to refer to the permanent gases—hydrogen and the cleveite gases. These appear in the stars which, on the grounds previously stated, I hold to be of highest temperature, and for this discrimination we can rely better on the cleveite gases than on hydrogen, for the reason that the latter is much more widely distributed *quâ* temperature. There are not many groups of stars which do not indicate the presence of hydrogen, while, on the other hand, the cleveite gases only occur conspicuously in one.

As pointed out in Chapter IV, it is impossible to over-estimate the importance of the discovery of terrestrial sources of helium in its bearings upon the spectra of the hotter stars, since it explains many of the strongest lines in such spectra, which previously were included in the category of "unknown lines."

Increasing Temperature.

If the varying intensities of the cleveite lines are studied in the case of the stars which have so far been considered, it is

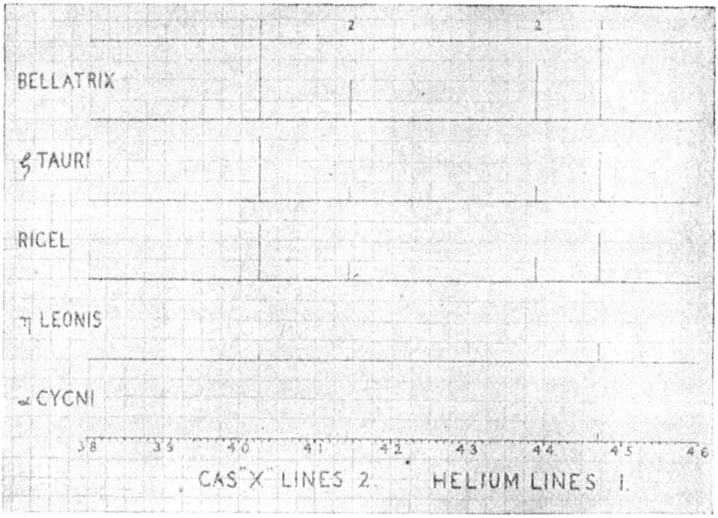

FIG. 77.—Behaviour of lines of cleveite gases in stars of increasing temperature.

found that the lines become stronger as the temperature increases. Thus, from the merest trace in α Cygni the lines are gradually intensified in passing through η Leonis, Rigel, ζ Tauri, and Bellatrix. This is illustrated in Fig. 77.

The lines of hydrogen make their appearance at a much lower stage of temperature than those of the cleveite gases, but, like the helium lines, they increase in intensity up to the highest temperature in the case of the stars already discussed.

So much being premised, we now come to a comparison of

the metallic results with those given by the cleveite gases and hydrogen in stars of ascending temperature.

Dealing with the stars already considered, a comparison of the metallic and cleveite gas lines indicates that as the former die out the latter are strengthened.

The cleveite gas absorption first makes its appearance, very feebly, in the stars in which the enhanced lines of magnesium and iron are strongest, as in *α* Cygni. Then, as the temperature increases, as demonstrated by the expansion into the ultra-violet of the continuous radiation, the lines of the cleveite gases become stronger as the metallic lines thin out.

At the highest temperature, taking Bellatrix for the present as a typical case, the principal helium lines are almost as strong as the lines of hydrogen, while the enhanced lines of iron have quite disappeared, and those of magnesium and calcium nearly so. These variations are indicated in Fig. 77.

It will be seen from the map that there is a perfect continuity both in the case of the metallic and the cleveite gas lines, and that in those stars where the separate investigations of the metallic and gas lines overlap, the same arrangement of stars results in both cases, *but there is a complete inversion in the behaviour of the gas, as compared with the metallic, lines.* We seem to be in presence of a chemical change, iron being finally replaced by helium.

The dark lines of hydrogen make their appearance in stars of comparatively low temperature, such as *α* Orionis, in which the second stage metallic lines are very strong. In these stars the hydrogen lines are thinner than some of the metallic ones.

With increased temperature, in the case of the stars so far considered, the lines of hydrogen steadily increase in intensity up to the highest temperature, as typified by Bellatrix. Meanwhile, as we have seen, the second stage metallic lines gradually thin out, while the enhanced lines become stronger up to the temperature of *α* Cygni, and ultimately all the lines of iron have disappeared, and magnesium and calcium are only represented by traces of their enhanced lines.

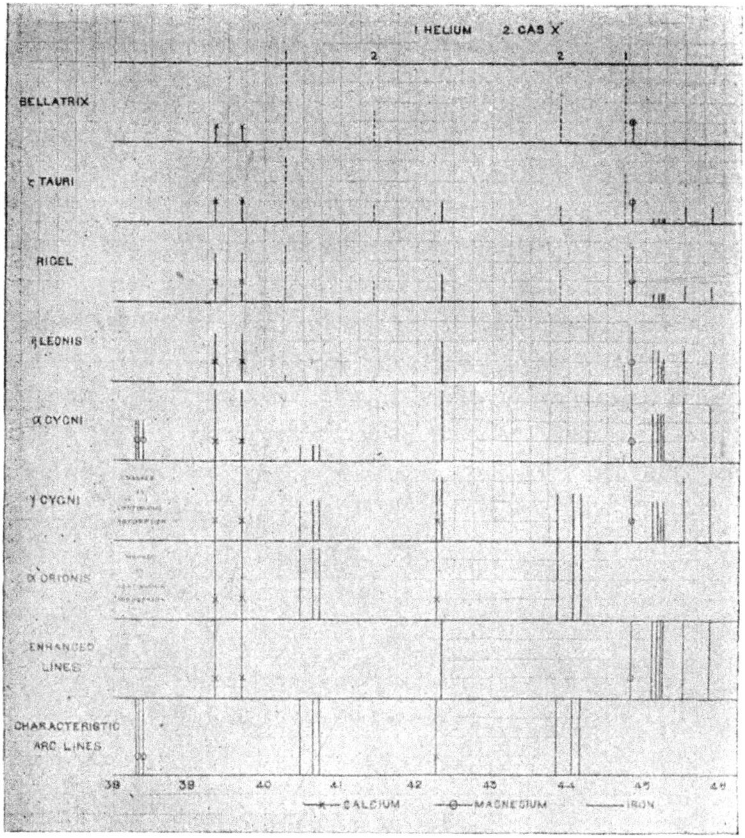

Fig. 78.—Comparison of metallic lines with lines of cleveite gases in stars of increasing temperature.

When the lines of the cleveite gases first become visible, as in *a* Cygni, the lines of hydrogen have already become the strongest in the spectrum; but at still higher temperatures the intensification of the lines of hydrogen is less rapid than that of the lines of the cleveite gases. In none of the stars now considered do the hydrogen lines reach their maximum development, as stated before. This point will be considered later on.

Descending Temperature.

We next consider the question in our list of cooling stars.

The appearance of the lines of the cleveite gases in these stars is restricted to β Persei and γ Lyræ. An attempt is made to show their changes of intensity in Fig. 79, which also includes Bellatrix as representative, for the present, of the highest temperature.

Starting with Bellatrix, the lines of the cleveite gases are very strong, while, on passing to the cooler stars, they rapidly become weaker, as the enhanced metallic lines continue to get stronger. In fact, the lines of the cleveite gases disappear at a comparatively early stage, long before the absorbing metallic vapours are cool enough to show any traces of the characteristic arc lines.

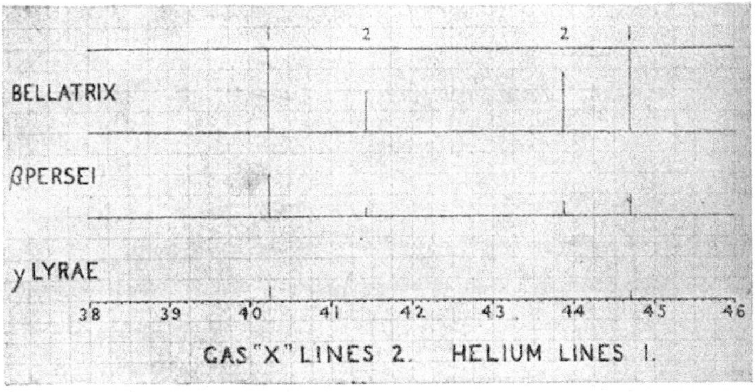

FIG. 79.—Behaviour of lines of cleveite gases in stars of decreasing temperature.

In β Persei, as shown in Fig. 79, the cleveite gas lines are much weaker than in Bellatrix, while in γ Lyræ they are on the verge of disappearance. No certain indications of the cleveite gas lines have been found in Sirius, Castor, Procyon, or Arcturus. A special search for the D^3 line in the spectrum of Sirius was made by Mr. Fowler in December, 1893, but the line was not seen in this star although it was recorded in the spectrum of Bellatrix.[1]

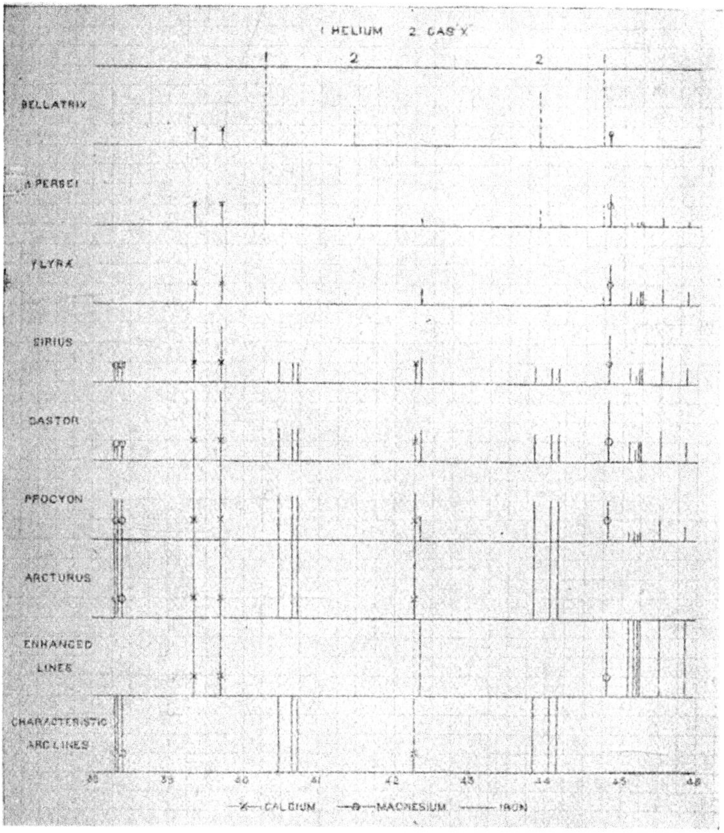

FIG. 80.—Comparison of metallic lines with lines of cleveite gases in stars of decreasing temperature.

[1] *Phil. Trans.*, vol. clxxxvi, A, p. 85.

So far as they go, then, the gradually disappearing lines of the cleveite gases indicate the same order of temperature as that determined from the gradual appearance of the metallic lines as shown in Fig. 76. Fig. 80 gives a comparison of the metallic and gas results in stars of decreasing temperature.

CHAPTER XIX.—APPLICATION OF THE NEW CRITERIA.

THE new classification which I suggested in 1888 was based on eye observations. It indicated that stars should be classed in two categories, one of ascending the other of descending temperatures. The photographic survey enabled us to utilise the length of the spectrum in the ultra-violet and violet as a preliminary basis of classification; the test was applied, and precisely the same order was arrived at as that obtained from the eye observations.

The general result of the test photographic survey, so far as it has gone, is as follows:—Among the 171 stars considered there are really two series of spectra, one representing the changes accompanying increase of temperature, while the other represents the effects of decreasing temperature. The fundamental requirement of the meteoritic hypothesis is, therefore fully justified by the discussion of the photographs.

It may be stated further that the sequence determined from the photographs follows exactly the same order as the groups originally suggested by the hypothesis from a discussion of the eye observations. That is, it is not necessary to interchange any of the groups founded on the eye observations in order to obtain agreement with the photographic results.

Finally, the enhanced-line work was undertaken, and then it was found that the order obtained from a consideration of the lines of iron or of calcium or of magnesium gave the same results as those gathered from the two previous enquiries; further, the identical behaviour of the enhanced metallic and cleveite lines on both sides of the curve shows us not only that we have here a previously undreamt-of test to apply to this question, but that the scheme of classification has borne the test in a most satisfactory manner, so far as temperature is concerned, up to the limit I have so far taken.

I have shown on page 309 that our highest terrestrial temperatures in all probability only carry us up to a temperature approximately represented by γ Cygni.

We have no right to assume that the small number of stars as yet studied put us in presence of the highest stellar temperatures; hence I have of set purpose left on one side for future treatment those stars which apparently are at the very apex of the temperature curve, for the reason that in these stars we are involved in unknown lines. These require a special study; other stellar photographs have to be examined, a work which will require some time, and new photographs have to be taken.

In all probability, among the stars already known there are others besides Bellatrix which are hotter than ζ Tauri and Rigel, as determined by the continuous spectrum and by the enhanced metallic lines in the way already explained; although the known gas lines are of nearly similar intensity, these stars show distinct differences among themselves, and other criteria for their arrangement in order of temperature must, therefore, be looked for.

At the stage represented by Bellatrix the only remaining known metallic lines are probably magnesium 4481, and the K line of calcium; these, however, are so feeble that their variations are too much dependent on the quality of the photographs to be trustworthy criteria, and the same remark applies also to the measurement of the extension of the spectrum into the ultra-violet. There are, however several lines, as yet of unknown origin, which are strong in some of these stars and weaker in others, and the appearances of these afford useful criteria for classifying these stars. But, before we attempt to use these lines to see whether any higher temperature than that of Bellatrix is indicated, it is important to consider whether we are justified in regarding these unknown lines as gaseous instead of metallic. This question is now being studied, and until it is settled it is not wise to attempt to discuss the upper temperature limits of those substances, such as hydrogen and the cleveite gases, the lines of which increase in importance with the increase of stellar temperature in the stars I have so far discussed.

If such hotter stars were found we might expect that either the hydrogen- or the cleveite-gas lines would thin out and then disappear first, as the iron lines thin out in Rigel and disappear in Bellatrix; and that they would be replaced by other lines. As a matter of fact, the hottest stars under discussion do contain other lines besides those of hydrogen and the cleveite gases with so far unknown origins.

Temperature is the fundamental point, but it is not the only one; other considerations must not be neglected; but, for the present, considering temperature alone and utilising the new criteria which have now become available, some of the brighter stars may be arranged as follows, those shown on the same horizon being of equal temperature, as determined by the enhanced metallic lines.

These criteria, however, do not enable us to classify minutely the stars which fall near the top of the curve, so, for the present, a certain number, which must be afterwards separated, are grouped together.

The stars which have been included in the maps (Figs. 74—80) are shown in italics.

Increasing Temperature.	*Decreasing Temperature.*
(*Bellatrix*, ζ Orionis, η Ursæ Majoris, λ Tauri, γ Pegasi).	
ζ *Tauri.*	β *Persei*, δ Cygni, α Pegasi, α Andromedæ, α Coronæ, γ Ursæ Majoris.
Rigel, β Tauri. η *Leonis*, α *Cygni.*	γ *Lyræ.* α Canum Venaticorum.
	Sirius, Vega, γ Geminorum, δ Leonis, β Ursæ Majoris, ε Ursæ Majoris.
γ *Cygni*, δ Cephei, Polaris, α Persei, ζ Geminorum.	*Castor*, β Arietis, α Cephei, α Aquilæ, δ Cassiopeiæ, β Cassiopeiæ.
	Procyon.
α Tauri, ε Pegasi, γ Andromedæ, ε Virginis.	*Arcturus* (Sun), α Arietis, β Geminorum.
α *Orionis*, β Pegasi.	

I have already indicated in Chapter XV how, on the meteoritic hypothesis, the physical conditions on either side of the curve must differ in stars of the same, or nearly the same, mean temperature. Indeed, I may generalise still more than I did then, and point out that on that hypothesis we must recognise three chief periods in the history of a star during its stages of luminosity:—

1. A period during which it exists as an uncondensed swarm when the "atmosphere" is disturbed by meteoritic bombardment from without. At this stage the "atmosphere" is a mass of heterogeneous vapour at various temperatures, and moving with different velocities in different regions.

2. A period of complete vaporisation, during which the atmosphere in quiescent, bombardment from without having ceased, and the radiation being too great to permit condensation in the upper parts of the atmosphere.

3. A period of cooling, during which the atmosphere is dis-

turbed by the fall of condensation products from the outer parts of the atmosphere on to the photosphere.

It is next important, therefore, to bring together the actual facts shown in the photographs of the spectra of those stars shown to be at about the same mean temperature on the two sides by the principles I have indicated.

With the photographs at present available, the test can be applied at three stages of temperature, the stars of equal mean temperature being determined by the relative intensities of the enhanced and arc metallic lines. When several stars of the same mean temperature are thus brought together, it soon becomes evident that they are divisible into two groups, which differ considerably in other respects. To take one instance, the average temperature of the absorbing iron vapour is about the same in ϵ Pegasi as in Arcturus, since the spectroscopic difference between these stars, so far as the line spectra are concerned, is very slight. But the continuous absorption in the violet is much greater in ϵ Pegasi than in Arcturus, while the metallic lines are also somewhat broader. The difference between the stars, therefore, does not appear to be due alone to a difference of temperature. Applying this method of separation, some of the typical stars given on the maps exhibit the following characteristics, those of equal temperature, as determined by the metallic lines, being placed on the same horizon.

Ascending Arm.	*Descending Arm.*
Rigel :—	γ *Lyræ* :—
1. Long continuous spectrum.	1. Continuous spectrum probably a little longer than in Rigel.
2. Hydrogen lines moderately thick.	2. Hydrogen lines very thick.
3. Metallic lines of moderate intensity and thickness.	3. Metallic lines weak and thin.
4. Cleveite gas lines of moderate intensity.	4. Cleveite gas lines very weak.
γ *Cygni* :—	*Castor* :—
1. Considerable continuous absorption in ultra-violet.	1. Very little continuous absorption in ultra-violet.
2. Hydrogen lines relatively thin.	2. Hydrogen lines relatively very thick.
3. Metallic lines of moderate intensity.	3. Metallic lines relatively feeble.
ε *Pegasi* :—	*Arcturus* :—
1. Strong continuous absorption in violet.	1. Little continuous absorption in violet.
2. Metallic lines thick.	2. Metallic lines of moderate thickness.

The facts thus indicate that at each of three very distinct stages of temperature there are two groups of stars showing spectroscopic differences. Generalising from these facts, it may be stated that stars at about the same temperature as judged by the iron lines, on the ascending side of the curve, differ from those on the descending side:—

(1) In the greater continuous absorption in the violet or ultra-violet, especially at the lower stages of temperature.

(2) In the relative thinness of the hydrogen lines at the higher stages of temperature.

(3) In the greater intensity and thickness of the metallic lines, whether of low or high temperature.

(4) In the relatively greater thickness of the lines of the cleveite gases at those stages of temperature in which they appear.

I submit that the differences found are precisely those we

should expect on the meteoritic hypothesis. I deal with the differences indicated *seriatim*.

1. *The Inequality of the Continuous Absorption.*—I have already pointed out that in the case of a *swarm* there must always be cooler vapours mixed with the hotter ones in the most valid absorbing regions. It is these cooler vapours which produce the absorption in the violet and ultra-violet.

In a *condensed mass of vapour* they can only exist at the limit of the atmosphere where their absorption is reduced in consequence of the low pressure.

2. *The smaller thickness of the Hydrogen Lines.*—The difference in thickness of the hydrogen lines is also sufficiently accounted for by the difference of absorbing conditions. In the stars of increasing temperature, consisting of uncondensed meteoritic swarms, the interspaces will be largely occupied by hydrogen at a high temperature, and the radiation of this gas will tend to mask the absorption produced by that in the immediate neighbourhood of the incandescent meteorites which merely graze. In the condensed stars with photospheres, any masking effect of this kind must be very much less pronounced, so that the hydrogen lines will be broader than in the spectra of uncondensed swarms. The hydrogen lines are more constant, in regard to their intensities, than the lines of other substances in the stars of increasing temperature, and this greater life of the hydrogen molecule seems to explain the fact that the hydrogen lines are specially picked out after passing to the downward side of the temperature curve when a state of quiescence is reached.

It should be remarked that the great distension of the hydrogen lines in the hotter stars which have begun to cool does not necessarily indicate a great thickness of absorbing hydrogen, since, in the case of the sun, the very broad H and K lines are produced by an absorbing region of small thickness in comparison with the sun's diameter.

3. *The greater thickness of the Metallic Lines.*—In the case of a swarm, the thickness of the effective absorbing gases and vapours in the line of sight will be very much greater than the effective thickness in the case of an atmosphere surrounding a photosphere, even if the masses and average temperatures be the same in both cases. The light proceeding from the central parts of the swarm must pass through the whole depth of vapours filling the interspaces, and on this account the absorption lines would be more intense in the case of a swarm. The greater intensities, and to some extent the breadth of the metallic lines, are thus explained, for since the metallic vapours will not fill the interspaces to the same degree that the hydrogen does, there will be no masking of the dark lines by radiation.

4. *The Widening and Thinning of certain Lines.*—In consequence of the great difference of velocity and direction of the meteorites entering a swarm, the spectrum lines involved will in general be broader, and therefore dimmer, so far as this cause is concerned, than in stars where bombardment has ceased. Such a broadening was specially noticeable in the spectrum of Nova Aurigæ, as I pointed out on page 229.[1]

The conditions, however, may vary considerably in different cases, according to the character of the parent nebula. In the case of a swarm which originally had a spiral structure, the chief movement will be very similar to one of rotation; if the axis of rotation be directed towards the earth, such a movement will produce little or no effect on the widths of the lines, but if the axis be not so placed, different amounts of broadening would be produced, according to the inclination and radial velocity. In these cases, lines which would be sharp at the edges, when there was no movement, would remain sharp at the edges when broadened and dimmed.

---

[1] *Proc. Roy. Soc.*, vol. l, p. 434.

When the original nebula was of less regular form, the influx of meteorites towards the centre will take place in a greater variety of directions, so that the broadening effects will be less regular.

Actions of this kind, in addition to those already referred to in 3, are probably to some extent responsible for the generally greater breadth of the metallic lines in bodies of increasing temperature as compared with those in cooling stars of the same mean temperature.

ζ Tauri is an interesting case in point. While the lines of hydrogen are quite sharp and not very broad, those of the cleveite gases are greatly distended and relatively dim. In this case, therefore, it would appear that the cleveite gases are more involved than hydrogen in the highest temperature collisions. In β Lyræ, also, the bright lines of the cleveite gases are more intense than those of hydrogen, and here we have another indication that these gases are among those chiefly involved in the spectral phenomena at the highest temperature; further, there is direct evidence that there are at least two bodies in the system of β Lyræ, and the variability is probably in part due to collisions between the outlying meteorites.

It may, on the other hand, be that the lines in the spectra of some cooling stars may be broadened as an effect of rotation, as suggested by Professor Pickering in the case of α Aquilæ. My own photographs show that the spectrum of this star is almost identical with that of β Arietis, except that all the lines are broadened. In this and similar spectra, such as α Ophiuchi and α Cephei, the broadening of the lines is accompanied by a reduction in intensity as in ζ Tauri.

5. *The greater thickness of the Lines of the Cleveite Gases.*—The action which produces the lines of the cleveite gases, whatever it be, only commences shortly before the highest temperature is reached, and the importance of helium in the spectrum grows very quickly. When the action has ceased, the helium lines

rapidly lose their importance, whilst the absorbing hydrogen continues for some time to become more effective. The complete discussion of these differences cannot be undertaken until the criteria for stars at the apex of the temperature curve have been further investigated.

CHAPTER XX.—A TEST CASE BETWEEN THE RIVAL HYPOTHESES.

AT the end of Chapter XVI it was pointed out that our knowledge of the spectra of stars of intermediate temperature was incomplete, and that more work was necessary. In the two Chapters preceding this some results of one of the inquiries set on foot to fill up the gap in our knowledge were indicated. I have now to call attention to a very special research into the minute details of the spectra of a few stars of intermediate temperature. The upshot of it is all the more interesting since the research was not undertaken in the first instance with the object of advancing the subject of classification, but rather to throw light on the phenomena of a very interesting class of variable stars, among them δ Cephei.

Along with δ Cephei the spectra of η Aquilæ, ζ Geminorum, T Vulpeculæ, and S Sagittæ were studied, and I pointed out[1] that judging by their spectra these belonged to the same sub-group as γ Cygni, which, however, is not a variable star; α Ursæ Minoris has also a spectrum which has since been recognised to be identical with that of γ Cygni and δ Cephei, while α Persei differs so slightly from them that it may also be classified with them. These I now group together, and for convenience of reference I designate them the δ Cephei class.

A portion of the spectrum of δ Cephei, at the time of maximum, is shown in Fig. 81, where it is compared with the spectra

[1] *Proc. Roy. Soc.*, vol. lix, p. 103.

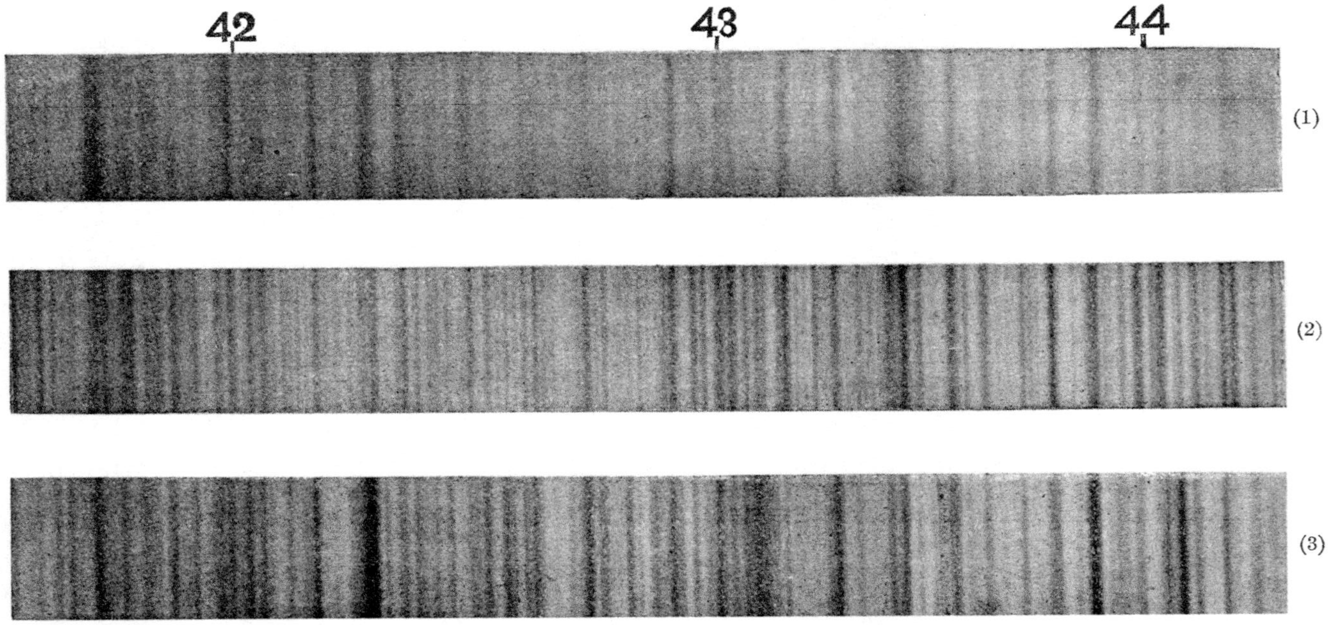

FIG. 81.—Comparison of the photographic spectra of δ Cephei (1), γ Cygni (2), and Arcturus (3).

of γ Cygni and Arcturus. These have been enlarged about ten times from the original negatives taken at Kensington.

Taking Arcturus as a representative star of the solar type,[1] it will be seen that although the spectra of γ Cygni and δ Cephei resemble it in showing a large number of dark lines, they differ considerably from it in point of detail.

Many of the lines in the spectrum of δ Cephei coincide with prominent lines in the spectrum of α Cygni, but it is true that many also coincide with lines in the spectra of stars like α Orionis and Arcturus, which closely resemble the solar spectrum. It seemed possible, therefore, that we might be dealing with the integrated spectra of two stars in close proximity, one having lines resembling those of α Orionis or Arcturus, and the other those of α Cygni. The spectra of all the stars of this sub-group have accordingly been very carefully investigated from this point of view. Enlarged glass positives of α Cygni and Arcturus on exactly the same scale have been superposed, and the integrated spectra photographed. When this integrated spectrum is compared with γ Cygni or δ Cephei, there is a considerable similarity, but the relative intensities of the various lines and the general appearance of certain parts of the spectrum, especially about G, are quite different. Again, if there were two bright bodies physically connected in such a star as γ Cygni there must be a revolution and a consequent doubling of the common lines, unless the plane of movement were perpendicular to the line of sight. No signs of such doubling, however, have been detected in any of the eight stars of the sub-group which have so far been recognised, and it is quite improbable that the plane of revolution would be at right angles to the line of sight in every case, and still more so that the two components would have identical spectra in each of the eight systems.

---

[1] *Phil. Trans.*, 1893, vol. clxxxiv, A, p. 699.

Spectra of the δ Cephei type must, therefore, be taken to represent a particular stage in the orderly development of cosmical bodies.

The spectra of the stars of the group, as will be seen from an inspection of the photograph, are distinguished by special characteristics. While containing a great number of fine metallic lines, giving them more or less the same general appearance as the solar spectrum, they show many lines which are either faint in the solar spectrum or are altogether absent; the spectra are practically identical with that of γ Cygni, which my previous work had indicated to be a star of increasing temperature.

The chief argument in favour of placing γ Cygni on the ascending side of the temperature curve was based on the presence of certain special lines, which occur with increased importance in the spectrum of α Cygni, the spectrum of which differs very widely from that of the sun, and has a close relationship to those of the Orion stars.

Further, the association of a special kind of variability with some of the stars having a spectrum of this type seemed to strengthen the view that the constitution of such stars must be vastly different from that of the sun.

Dr. Vogel, however, has classified two stars of the group, namely, η Aquilæ and S Sagittæ, with the sun in his Class IIa, and more recently Dr. Scheiner has placed another star of the group (α Persei) between α Cygni and the sun. I must also add that Drs. Vogel and Scheiner differ very considerably as to the classification of α Cygni, and this still further complicates matters.

It will be seen that the question between the two classifications is a very sharp one.

In Vogel's classification all stars are regarded as cooling bodies, while one of the chief points of mine is the distinction between stars which are getting hotter and those which are becoming cooler. Thus, while stars like δ Cephei and those like the sun are grouped together by Vogel in his Class II*a*, they are in mine divided into two groups, Group III including δ Cephei and Group V the sun. The important question to be answered may be formulated as follows:—

Is the difference between stars like δ Cephei and stars resembling the sun solely due to a small temperature difference, as it is on Vogel's view, or does it in part represent, as I contend, a physical difference? On Vogel's view, the stars in question are like the sun, while on mine they are not. To make the difference quite clear I repeat that I hold that stars like δ Cephei consist of uncondensed swarms of meteorites of increasing temperature, while those like the sun are cooling bodies and consist of masses of vapour in which there are photospheres and relatively quiet atmospheres.

Several questions connected with the variability and light curve of these stars, and especially some new observations of η Aquilæ, one of the group, induced me to give special attention to them, from the point of view of their classification. I propose therefore to refer to the new work, which has been as detailed as possible, in order to arrive at a just conclusion. Such work as this, indeed, must be detailed to be worth anything.

But before I state the results at which I have arrived, it is necessary, in order to enable a judgment to be formed between Dr. Vogel's views and my own, to refer to the location of this group of stars in his classification with equal detail.

DR. VOGEL'S CLASSIFICATION.

η Aquilæ and S(10) Sagittæ are the only two of the stars which have been mentioned as belonging to the δ Cephei class included in Vogel's spectroscopic *Durchmusterung*, published in 1883; and both are classed without further comment as stars of Class II*a*.[1] This type of spectrum was thus defined:—

[1] *Public. Astr. Obs. zu Potsdam*, vol. iii, p. 200.

"Spectrum with very numerous metallic lines, which are easily known by their intensity, especially in the yellow and green. The hydrogen lines are for the most part strong, but are never so broad as in the case of Class I*a*. In some stars the lines of hydrogen are faint, and in these faint bands can be generally recognised in the less refrangible portion of the spectrum."

Although the other stars are not included, it seems fair to suppose that all the stars of the δ Cephei class would be classed with the sun if they had been included.

Dr. Scheiner has discussed one of the stars of the δ Cephei class, α Persei, in some detail.[1] Attention is specially drawn by him to the differences between the spectrum of α Persei and that of the sun.

In the table on page 335 I have brought together the lines of α Persei which Dr. Scheiner states to be more intense than in the sun, and have compared them with those of δ Cephei and α Cygni photographed at Kensington. The remarks in the last column are those made by Dr. Scheiner with regard to the spectrum of α Persei.

It will be seen that the special lines of α Persei are nearly all present in δ Cephei and α Cygni, and that they are all either faint in, or absent from, the solar spectrum.

Dr. Scheiner also remarks on a few of the special lines of α Ursæ Minoris, another member of the δ Cephei class. My own photographs show almost absolute identity with δ Cephei, so that a special discussion of this star would be superfluous.

The spectrum of δ Cephei has been investigated by Belopolsky,[2] and he gives a table showing that it differs in many respects from that of the sun. Attention is specially drawn by him to some of the lines which are strongly marked in the spectrum of δ Cephei as compared with corresponding lines in the solar spectrum photographed with the same instrument;

---

[1] *Public. Astr. Obs. zu Potsdam*, vol. vii, Part II, p. 329.

[2] *Ast. Nach.*, No. 3338, p. 19.

Special Lines in α Persei.

α Persei (Scheiner), λ Potsdam.	δ Cephei (Lockyer), λ Rowland. Maximum intensity = 10.	α Cygni (Lockyer), λ Rowland. Maximum intensity = 10.	Remarks on α Persei (Scheiner).
4290·2	4290·7 (8)	4290·9 (4)	Stronger than in sun.
4306·3	4306·4 (5)	4306·3 (1)	,, ,,
4310·1	..	..	No corresponding solar line.
4313·4	4313·3 (4)	4313·7	Stronger than in sun.
4321·4	?	4321·4 (2)	Much stronger than in sun.
4331·0	4331·1 (6)	4331·0 (2)	Stronger than in sun.
4344·7	4344·5 (5)	4344·0 (2)	,, ,,
4375·1	4375·0 (10)	4374·7 (2)	Much stronger than in sun.
4387·2	?	4387·8 (1)	Stronger than in sun.
4391·4	4391·4 (5)	4390·9 (4)	,, ,,
4394·4	4394·4 (6)	4395·2 (6)	,, ,,
4400·0—4400·9	Broad line (8)	4400·1 (5)	,, ,,
4411·1	4411·0 (4)	..	No solar line.
4413·6	4413·6 (1)	..	,, ,,
4416·9	..	4417·0 (7)	,, ,,
4450·7	4451·1 (5)	4451·1 (3)	Stronger than in sun.
4461·7	4462·0 (5)	4462·0 (2)	,, ,,
4464·7	4465·0 (2)	4464·1 (2)	,, ,,
4468·7	4468·4 (6)	4468·7 (5)	,, ,,
4471·1	4471·1 (3)	4471·3 (3)	No solar line.
4473·1	4473·1 (3)	4473·2 (2)	Stronger than in sun.
4481·6	4481·3 (7)	4481·3 (10)	,, ,,
4488·6	4489·3 (5)	4489·1 (5)	,, ,,
4491·8	4492·0 (3)	4491·5 (4)	,, ,,
4501·5	4501·5 (7)	4501·6 (4)	Much stronger than in sun.
4508·5	4508·5 (5)	4508·5 (7)	,, ,,
4515·6	4515·6 (6)	4515·7 (6)	Stronger than in sun.
4520·6	4520·5 (3)	4520·4 (6)	,, ,,
4534·4	4534·3 (7)	4534·5 (6)	Much stronger than in sun.
4545·3	4545·3 (6)	4545·4 (1)	,, ,,
4549·9	4549·9 (7)	4549·9 (8)	Stronger than in sun.
4564·1	4564·1 (4)	4564·2 (5)	,, ,,
4572·0	4572·0 (5)	4572·5 (5)	,, ,,

a discussion of these differences shows that the more special lines of δ Cephei, like those of *α* Persei, are very prominent in *α* Cygni. In spite of these differences, however, Dr. Scheiner classes *α* Persei with stars like the sun, but regards it as a transition stage between *α* Cygni and the sun. He says :—

"From the general conclusions, the above list gives us quite a curious and important result. If one compares these lines with those in the spectrum of α Cygni, which is of special interest as an advanced spectrum of Class I*b*, one finds that, out of sixty lines in α Cygni, twenty-one occur in the list, and all of these in α Persei. The fact, therefore, remains that nearly half the lines which in the spectrum of α Persei show divergences when compared with the solar spectrum appear certainly in any case in the spectrum of α Cygni and determine its peculiarity.

"Now α Cygni belongs to Class I*b*, α Persei to Class II*a*, and the plausible conclusion appears to me, therefore, that α Persei had previously a spectrum similar to that of α Cygni, and that in this case we have the missing connection between I*b* to II*a*." [1]

The spectrum of α Tauri, which, as I shall show, must be regarded as a condensing swarm, closely resembles that of the sun; so that Dr. Scheiner's evidence would equally place α Persei intermediate between α Tauri and α Cygni.

Difficulties connected with Vogel's Classification.

The close association of stars like δ Cephei with those like α Cygni makes it here important to discuss the place of α Cygni in Vogel's classification. With regard to this star Dr. Scheiner writes:—

"The spectrum of α Cygni, in spite of the large number of its lines, has no resemblance with that of the sun. While it is possible to identify most of the lines with solar lines in respect of their position, yet the total lack of agreement as to intensity of the lines makes many of these identifications worthless." [2]

Dr. Scheiner has classified α Cygni in Vogel's class I*b*, which was thus defined in Vogel's original classification: "Spectra in which the metallic lines are few in number, and very faint or entirely imperceptible, and in which the hydrogen lines are lacking."

In view of the photographic results obtained at Potsdam, the last clause in this definition was revised in 1888, so that it

---

[1] *Potsdam Observations*, vol. vii, Part II, p. 331.

[2] Scheiner's *Astronomical Spectroscopy* (Frost's translation), p. 247.

reads: "and the strong hydrogen lines of Type I*a* are lacking."[1] This has again been modified by Dr. Scheiner, and the characteristics of Type I*b* are stated as "spectra in which the hydrogen lines and the few metallic lines all appear to be of equal breadth and of sharp definition."[2] It is on the ground of this greatly modified definition that *a* Cygni is included with stars like Rigel in Class I*b*.

Dr. Vogel, however, is not prepared to accept Dr. Scheiner's amended definition of Class I*b*. He writes:—

"However justifiable it may be to regard the peculiarly sharp spectral lines of the stars above mentioned and a few others of the same kind as worthy of special consideration, the adoption of this proposal would make it necessary to separate a number of stars (including those of Orion) whose relationship is placed beyond all question by the investigations I have referred to, and to place them with *a* Cygni, which has a materially different spectrum."[3]

In the same paper Dr. Vogel brings forward a new definition of his Class I*b* in the following terms:—

"Spectra in which, besides the still dominant hydrogen lines, the lines of cleveite gas appear, and above all the lines λ 4026, λ 4472, λ 5016, and λ 5876(D_3). The lines of calcium, magnesium, sodium, and iron are also more or less numerous in spectra of this subdivision."

This new definition excludes *a* Cygni from Class I*b*, in Dr. Vogel's opinion, and he places this star in Class I*a* 3 of his extended classification of spectra of the first class. This subclass is thus defined:—

"Spectra in which the calcium line λ 3934 has nearly the same intensity as the hydrogen lines. In occasional instances it is still sharply defined at the edges, or it may be broader and more intense than the hydrogen lines, and very diffuse, forming with the hydrogen line H_ϵ(λ 3970), which is greatly intensified and broadened by the calcium line λ 3969, a conspicuous pair.

"In the spectra of this division the lines of the cleveite gas cannot be recognised; on the other hand numerous strong lines of different metals,

---

[1] *Ast. Nach.*, vol. cxix, p. 97.

[2] Scheiner's *Astronomical Spectroscopy* (Frost's translation), p. 245.

[3] *Astrophysical Journal* (1895), vol. ii, p. 343.

particularly lines of iron, are always present. The lines of hydrogen are still always dominant. H_δ is plainly apparent among the other lines, and the group G is less conspicuous than H_γ. This subdivision forms the direct connecting link with the spectral Class II, in which the hydrogen lines no longer play a prominent part in comparison with the lines of other metals."[1]

According to this definition, α Cygni is classed with Procyon, a star which in the main resembles the sun, and this notwithstanding Scheiner's remark that the spectrum of α Cygni bears no resemblance to that of the sun.

It is clear, then, from these disagreements among the Potsdam observers, and the frequent changes made in the definition of the various classes, that it is difficult to classify α Cygni satisfactorily on Vogel's hypothesis that all stars are cooling.

My Own Classification.

The First Arguments.

I gave in the *Meteoritic Hypothesis* the evidence that in stars such as α Orionis, the occurrence of radiating carbon vapour is an indication that these stars, like comets, consist of uncondensed swarms of meteorites. Since the temperature of a condensing swarm of meteorites must be increasing, in accordance with thermodynamical principles, stars like α Orionis must be placed on the ascending arm of the temperature curve. The photographs show that the spectrum of stars like α Tauri is almost identical with that of α Orionis so far as the lines are concerned, and since one of the flutings in the red in the spectrum of α Orionis also appears in α Tauri, this star must also be regarded as one of increasing temperature.

The discussion of the Kensington photographs led me to place γ Cygni next to α Tauri in the series of stars with increasing temperature, and we now know that δ Cephei must

[1] *Astrophysical Journal* (1895), vol. ii, p. 344.

be classed with this star. I pointed out that the spectrum of γ Cygni

"Has much in common with that of α Tauri, but there is less continuous absorption, and many of the lines of α Tauri thin out. The next step to α Cygni is rather a long one, but it seems very probable that if more photographs were available intermediate spectra would be found. It will be seen, however, that in α Cygni the hydrogen lines are intensified as compared with γ Cygni, and that all the important lines of α Cygni agree in position with prominent lines in γ Cygni. . . . . . In passing to Rigel the more important lines of α Cygni are retained, and a few new lines make their appearance."

My argument was, that with an increase of temperature a star like α Tauri would develop into one like δ Cephei, which, with further increase, would pass through successive stages represented by α Cygni, Rigel, and Bellatrix. To justify this it is accordingly necessary to show greater reason for associating δ Cephei with α Tauri than with a star like the sun, which we know on other grounds to be cooling; and from the great similarity of the line spectra of α Tauri and the sun, it is clear that the argument must not entirely depend upon the identity of lines in the spectra of α Tauri and δ Cephei, but upon general and specific differences between δ Cephei and the sun.

The New Argument derived from the Study of the Enhanced Lines.

It is most fortunate that on all the foregoing points the study of the enhanced lines throws a flood of new light (see Fig. 82).

A study of the enhanced lines shows at once that δ Cephei is hotter than either α Tauri or the sun, and that the difference between its spectrum and that of α Tauri or of the sun is certainly in part due to this difference of temperature of the absorbing vapours. The lines which are stronger in δ Cephei than in the sun include many of those which have been found to be enhanced in the spark spectra of metals, so that they are no longer to be regarded as unknown lines. Similarly, many of the lines of α Cygni for which no origins could previously be assigned have been shown to be lines of common metals under conditions of high temperature. Still, the mere presence of the enhanced lines in a star spectrum affords us no criterion as to whether the temperature of a star is increasing or decreasing. But I have also indicated on page 322 that if we take the relative intensities of the enhanced lines and the arc lines as an indication of stellar temperatures, and in this way bring together a sufficient number of stars of about the same temperature as γ Cygni or δ Cephei, such spectra may be divided into two well-marked groups, of which γ Cygni and Castor may be taken as types. The chief generic differences between the two groups of stars at the temperature of δ Cephei were thus summarised :—

δ Cephei.	Castor.
(1) Considerable absorption in ultra-violet.	(1) Very little continuous absorption in ultra-violet.
(2) Hydrogen lines relatively thin.	(2) Hydrogen lines relatively very thick.
(3) Metallic lines of moderate intensity.	(3) Metallic lines relatively feeble.

It was further shown that these differences are simply and sufficiently explained on the supposition that stars like γ Cygni and δ Cephei are uncondensed swarms of meteorites, while those like Castor, which have about the same mean temperature, are stars approaching the condition of the sun, in which photospheres and relatively quiescent atmospheres have formed.

The Strengthening of the Argument from Variability.

The evidence in favour of placing stars like δ Cephei and γ Cygni on the ascending arm of the temperature curve by the fact that stars of this class present a special form of variability has recently been greatly strengthened. This variability is similar in kind, but different in degree, to that associated with

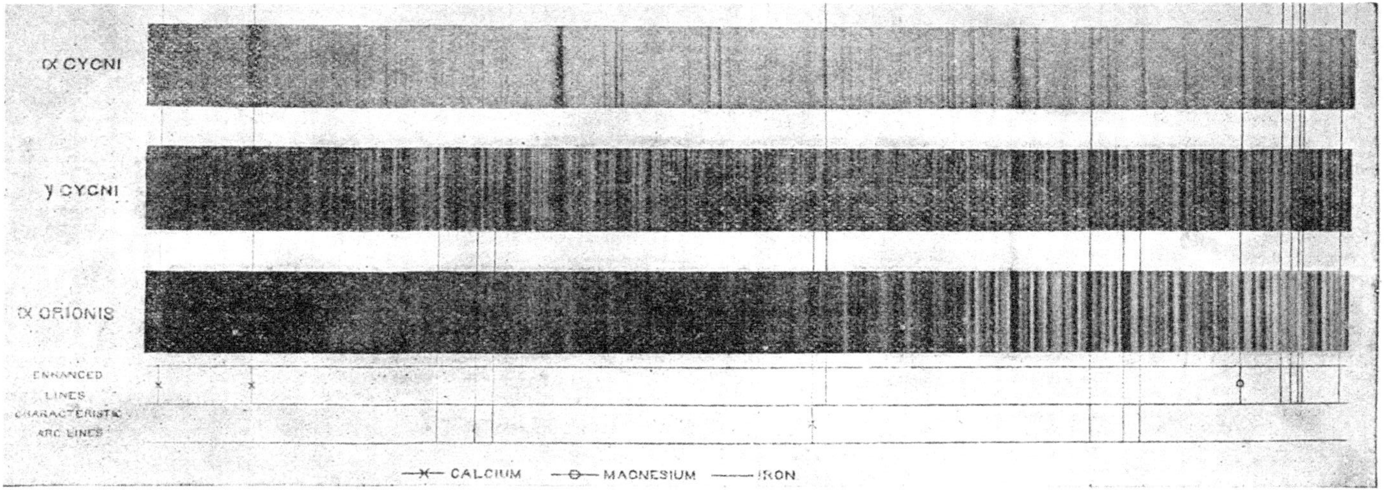

FIG. 82.—Untouched photographs of stellar spectra indicating how the enhanced lines are brought into the evidence. The photographs show that the triplets of the arc spectrum of iron are present in α Orionis and γ Cygni and absent from α Cygni, and that the quartet of enhanced iron lines is present in α Cygni and absent from γ Cygni and α Orionis.

stars of Group II, such as Mira. The following table shows that the amount of variation is very much less than that in variables of the Mira type :—

Name.	Period.		Interval from min. to max.		Variation.	Remarks.
	d.	h.	d.	h.		
δ Cephei......	5	9	1	15	3·7—4·9	Very regular.
ζ Geminorum .	10	4	5	0	3·7—4·5	Period slowly lengthening.
η Aquilæ......	7	4	2	9	3·5—4·7	Slightly irregular.
S(10)Sagittæ..	8	9	3	10	5·6—6·4	
T Vulpeculæ..	4	10	1	7	5·5—6·5	
Mira.........	331	0	..		(1·7—5·0) — (8·7—9·5)	

It will be seen that the luminosity at maximum is from about two (0·8 magnitude) to three times (1·2 magnitudes) greater than at minimum, while the forms of most of the light curves resemble the majority of those of the Mira class in the relatively steep ascent to maximum. A constitution more or less similar to that of the Mira class is therefore indicated.

I have already shown that in such variables as Mira the presence of bright carbon flutings indicates a meteoritic structure.[1] Here the variation has a much longer period than in δ Cephei, but it is only necessary to suppose that δ Cephei is more condensed, so that revolving swarms of short period will be alone effective in producing collisions at periastron, as I pointed out in 1889.[2]

A recent discussion of all the available observations of η Aquilæ by Dr. William J. S. Lockyer[3] has shown that the light curve of this variable can be best explained on the supposition of three meteor swarms moving around their centre of

[1] *Proc. Roy. Soc.* (1887), vol. xliii, p. 130.

[2] *Proc. Roy. Soc.*, vol. xlvi, p. 420.

[3] *Resultate aus den Beobachtungen des veränderlichen Sternes η Aquilæ* (Inaugural Dissertation), Göttingen, 1897.

gravity. In this way not only is the general form of the light curve satisfied, but the smaller irregularities discovered by the discussion are also easily accounted for.

Hence, by placing stars of the δ Cephei class on the ascending arm of the temperature curve, the variability of certain members of the group finds a ready explanation. I am not aware of any satisfactory explanation of the δ Cephei type of variability in which a constitution resembling that of the sun is assumed, and to my mind such a variation in a star constituted like the sun is impossible.

The final result of the discussion of the spectra of stars of the δ Cephei class is to show that they must be placed on the ascending arm of the temperature curve, at a stage higher than stars like *α* Tauri, in which the mean temperature is not very different from that of the sun. Stars of equal temperature on the descending side of the curve, of which Castor may be taken as a type, show precisely the same lines, the enhanced and cool lines having the same relative intensities, but with inverted intensities of the hydrogen and metallic lines, and with somewhat less continuous absorption in the ultra-violet. The difference between stars like δ Cephei and those of the sun is therefore partly due to a difference of temperature and partly due to a difference of physical condition such as is demanded by the meteoritic hypothesis. On this hypothesis, the variables of this class are to be regarded as meteor-swarms not yet completely condensed. This result enables us to understand why some members of the δ Cephei class should show such a very special kind of variability.

α Cygni also finds a natural place on the ascending arm of the temperature curve, at a stage higher than δ Cephei, and all the difficulties met with in attempting to classify it on Vogel's view of decreasing temperature are removed.

Conclusion.

I am not aware that any more crucial test than the foregoing can be applied to the rival schemes of stellar classification, and as I hold that the result of its application is entirely in favour of the one which assumes the existence of some stellar bodies which are increasing their temperature while others are reducing it, the Sun's place in Nature must be regarded as near that occupied by Arcturus and Capella, and very far separated from that occupied by α Cygni, γ Cygni, and α Tauri.

Nor is this all, the origin of the Sun in a nebula not exclusively gaseous but only containing gases among its constituents is greatly strengthened by the extended study of the classification problem which has occupied the last few chapters.

Along all lines, then, the fundamental requirements of the Meteoritic Hypothesis have been strengthened by the later work.

FINIS.

APPENDIX.

APPENDIX.

THE SPECTRA OF NOVÆ. (See p. 234.)

Orion Nebula.		Bright Line Stars.		Nova Aurigæ.			Nova Cygni.		
Lockyer.	Campbell.[1]	Pickering.[2]	Campbell.[3]	Lockyer.	Campbell.[4]	Vogel.[5]	Cornu.[6]	Copeland.[7]	Vogel.[8]
3707·4									
3715·4									
3729·5									
3743·6									
3752·6									
3770·6									
3796·6									
3832·6									
—	3835	..	..	..	..	3835			
3847·6									
3855·7									
3868·7	3869								
3887·7									
—	3889	389	..	..	..	3889			
3902·7									
3910·7									
3933·7	..	..	..	3933·7	..	3934			
3941·7									
3949·7	..	395							

3968·7	3969	..	..	3968·7	396	3969			
3984·7	..	398							
4000·7									
4010·7									
4025·7	4026	402	4023						
4041·6									
4054·6									
4067·7	4067	406	4063	..	..	4067			
4086·7									
—	..	..	..	..	4098				
4101·8	4102	410	4102	4101·8		4102			
4120·8	..	412							
—	..	..	..	..	..	4125			
4129·8	..	..	..	4128·8					
4142·7									
4154·7									
—	..	..	..	..	..	4158			
4167·7									
—	..	..	..	4172·6					
—	..	..	..	..	..	4176			
4204·6	..	420	420	4202·6					
4226·6	..	..	4228	4226·6					
4234·6	..	..	..	..	423	4230			
—	423								
—	..	..	4260	..	..	4262			
—	..	..	..	4264·6					
4269·6	4265	..	4273	..	426				
—	..	..	..	..	..	4288			
—	..	..	..	4291·6					

1 *Ast. and Ast. Phys.*, vol. xiii, 1894, p. 500.
2 *Astr. Nachr.*, No. 3025, 1891.
3 *Ast. and Ast. Phys.*, vol. xiii, 1894, p. 473.
4 *Ibid.*, vol. xii, 1893, p. 726.
5 *Ast. and Ast. Phys.*, vol. xii, 1893, p. 912.
6 *Comptes Rendus*, vol. lxxxiii, p. 172.
7 *Copernicus*, vol. ii, p. 111.
8 *Berlin Akad. Monatsber.*, 1877, p. 243.

THE SPECTRA OF NOVÆ—*continued*.

Orion Nebula.		Bright Line Stars.		Nova Aurigæ.			Nova Cygni.		
Lockyer.	Campbell.[1]	Pickering.[2]	Campbell.[3]	Lockyer.	Campbell.[4]	Vogel.[5]	Cornu.[6]	Copeland[7]	Vogel.[8]
—	..	..	..	4310·6					
—	..	..	4318	..	..	4315			
—	..	..	4334						
—	..	..	..	..	4336				
4340·6	4341	434	4341	4340 6	..	4341	..	..	(434)
—	..	..	..	..	4358	..	435		
—	4364								
—	..	..	4369						
—	..	..	..	..	..	..	..	4376	
4385·7	..	..	..	4383·7	438	4383			
4389·7	4389	..	4389						
4410·7	..	..	..	4412·7					
—	..	..	4416	..	..	4417			
4426·7									
—	..	443	..	4434·7	..	4435			
—	..	..	4442						
—	..	..	..	..	..	4445			
—	..	..	4457						
—	..	..	4466	..	4466				
—	..	..	..	4469·7					
4471·8	4472	447	4473	..	..	4473			
—	..	..	4480	..	..	4480			
—	..	..	4493	..	..	4495			
4495·8									
—	..	..	4504						

—	..	..	4509	..	..	4507			
—	..	451	4517	4518·8	451	4520	451		
—	..	..	..	..	..	4530			
4539·8	..	455	4541						
—	..	..	4555	4555·8	..	4557			
—	4574								
—	..	..	..	..	..	4583			
—	..	..	..	4587·8					
—	459	..	4596						
—	..	..	..	..	460				
—	4610								
—	..	..	4615						
—	..	..	4626	4625·8					
4627·8	..	..	..	..	..	4628			
—	..	..	..	..	4630				
—	..	..	4636	..	..	..	..	4635	
—	..	464							
—	..	..	4652	..	..	..	..	..	(467)
—	4663·0								
—	..	469	4688						
—	..	..	..	..	4681				
—	4714·0	..	..	..	471	..	..	..	470
4715·9									
4735·9	..	..	..	..	..	..	..	..	474
—	..	..	4787						
4825·0	..	..	..	..	..	..	483		
4840·0									
—	..	..	..	..	4857				
4862·0	4861	486	4862	486	..	4862	..	4861	486

[1] *Ast. and Ast. Phys.*, vol. xiii, 1894, p. 500.
[2] *Astr. Nachr.*, No. 3025, 1891.
[3] *Ast. and Ast. Phys.*, vol. xiii, 1894, p. 473.
[4] *Ibid.*, vol. xii, 1893, p. 726.
[5] *Ast. and Ast. Phys.*, vol. xii, 1893, p. 912.
[6] *Comptes Rendus*, vol. lxxxiii, p. 172.
[7] *Copernicus*, vol. ii, p. 111.
[8] *Berlin Akad. Monatsber.*, 1877, p. 243.

THE SPECTRA OF NOVÆ—*continued.*

Orion Nebula.		Bright Line Stars.		Nova Aurigæ.			Nova Cygni.		
Lockyer.	Campbell.[1]	Pickering.[2]	Campbell.[3]	Lockyer.	Campbell.[4]	Vogel.[5]	Cornu.[6]	Copeland.[7]	Vogel.[8]
4897·9									
—	..	..	..	490					
4923·9	..	..	..	..	..	4923			
—	..	..	..	.	..	4925			
—	..	..	4940						
—	..	..	..	..	4953				
4958·0	4958								
—	..	..	..	..	5002	..	500	500	499
5007·3	5007	..	..	5006					
—	..	..	..	..	..	5016			
—	..	..	5020	..	..	..	..	5023	
—	..	..	..	..	..	..	..	..	508
—	..	..	5131	..	..	..	..	..	512
—	..	..	..	..	..	..	..	..	514
—	..	..	..	..	..	5167	517		
—	..	..	..	518					
—	..	..	..	..	..	523			
—	..	..	5250						
—	..	..	5284	..	..	528	..	..	527
—	5313	..	..	531	..	5317	531		
—	5412	..	5412						
—	..	..	5472	..	..	..	(546)		
—	..	..	..	558					
—	..	..	5593						
—	..	..	..	566	..	..	(563)		

—	..	..	..	..	..	5686			
—	..	..	5693	570					
—	5751	..	..	..	5750				
—	..	..	..	579	..	..	..	5775	580
—	..	..	5813						
—	..	..	5848						
—	..	..	5877						
—	..	..	..	589	..	..	588	589·5	[588]
—	..	..	..	630					
—	..	..	..	635	..	..	(635)		
—	..	..	6564	656	..	6562	..	6562	
—	..	..	..	..	..	..	661	C	
—	..	..	..	..	..	705			

1 *Ast. and Ast. Phys.*, vol. xiii, 1894, p. 500.
2 *Astr. Nachr.*, No. 3025, 1891.
3 *Ast. and Ast. Phys.*, vol. xiii, 1894, p. 473.
4 *Ibid.*, vol. xii, 1893, p. 726.
5 *Ast. and Ast. Phys.*, vol. xii, 1893, p. 912.
6 *Comptes Rendus*, vol. lxxxiii, p. 172.
7 *Copernicus*, vol. ii, p. 111.
8 *Berlin Akad. Monatsber.*, 1877, p. 243.

INDEX.

D.

HARRISON AND SONS, PRINTERS IN ORDINARY TO HER MAJESTY, ST. MARTIN'S LANE.

Printed in Dunstable, United Kingdom

91219937R00218